T0418433

NEW TECHNOLOGY FOR CULTIVATING

Super Hybrid Rice

WITH A YIELD OF 900 KG/*MU*

www.royalcollins.com

NEW TECHNOLOGY FOR CULTIVATING

Super Hybrid Rice

WITH A YIELD OF 900 KG/*MU*

Chief Editors: Ma Guohui and Yuan Longping

Translator: Hu Dongping et al.

New Technology for Cultivating Super Hybrid Rice with a Yield of 900 kg/mu

Chief Editors: Ma Guohui and Yuan Longping
Deputy Chief Editors: Wei Zhongwei, Wu Chaohui, Long Jiyue, and Huang Sidi
Contributors: Li Jianwu, Huang Zhinong, Song Chunfang, Zhou Jing, and Wen Jihui
Translators: Hu Dongping, Li Yan, Li Jing, Yang Xuyun, Zheng Jingya, Li Tian, and Hu Tengyu
Proofreaders: Hu Tengyu and Hu Dongping

First published in 2025 by Royal Collins Publishing Group Inc.
Groupe Publication Royal Collins Inc.
550-555 boul. René-Lévesque O Montréal (Québec) H2Z1B1 Canada

Original edition © Hunan Science & Technology Press

All rights reserved. Without limiting the rights under copyright reserved above, no part of this publication may be reproduced, stored in or introduced into a retrieval system, or transmitted in any form or by any means (electronic, mechanical, photocopying, recording, or otherwise), without the prior written permission of both the copyright owner and the above publisher of this book.

ISBN: 978-1-4878-1304-8

To find out more about our publications, please visit www.royalcollins.com.

Contents

Preface

As food security is a major issue affecting the national economy and the people's livelihood, we are required to further strengthen our food production capacity per unit of land area to a much higher level. Rice is the most prominent food crop in China, and the continuous breakthrough in super rice yield is the most significant achievement with the rapid progress of science and technology in the rice research field in China. Hence achieving super-high rice yield is an important technical support to guarantee national food security.

Since the adoption of the China Super Rice Breeding Project by the Ministry of Agriculture and Rural Affairs in 1996, researchers in the rice research field have experienced more than 20 years of joint efforts to select and breed a number of super rice varieties that will make different yield potentials, creating the record with high yield one after another. A case in point is Yuan Longping's hybrid rice innovation team from the Hunan Hybrid Rice Research Center, which achieved the key goals of 100 *mu* with an average yield of 700 kg, 800 kg, 900 kg, and 1,000 kg per *mu* in 2000, 2004, 2012 and 2014, respectively, which contributed to China's leading role in the world in super hybrid rice breeding and cultivation technology.

In order to give full play to the technological achievements related to high yield of super hybrid rice, we convened relevant researchers who had participated in the research on high yield of super hybrid rice for years to jointly compile the book *New Cultivation Techniques for Super Hybrid Rice with a Yield of 900 kg/mu*, with new achievements in this field comprehensively summarized and collected, focusing on the characteristics of super hybrid rice varieties, cultivation technology to strengthen seedlings of super hybrid rice, fertilization technology for super-high yield of super hybrid rice, water management technology for super-high yield of super hybrid rice, the quality design and cultivation of super-high-yielding groups in super hybrid rice, the technology model of super hybrid rice with high-yielding and high efficiency, the evaluation of climate and ecological adaptability for super hybrid rice, typical case study of super hybrid rice with super-high yield, integrated preventing and controlling technique for major diseases and insect pests of super hybrid rice, etc.

Featured by its brevity and clarity, this book, undoubtedly easy to understand, is highly readable and operable, which can be used by researchers, agricultural technology extension personnel, and most rice farmers for reference in research and demonstration, popularization, and application. It is also a scientific, advanced, and practical manual on rice cultivation technology.

Due to time constraints and the authors' limited capabilities, some errors and omissions in this book are inevitable. It would be much appreciated if you could provide any constructive suggestions and appropriate feedback.

2021, August
Editor

CHAPTER 1

Overview

I. The Development Course of Hybrid Rice

Yuan Longping's research on hybrid rice dates back to 1964, closely followed by 1966, when he published his paper "Male Infertility in Rice" in the fourth issue of *Science Bulletin* based on his studies. He proposed breeding with "three lines" to exploit rice's heterosis advantage.

Under the leadership of Yuan Longping, the year 1973 witnessed the cooperation of "three lines" of hybrid rice, namely cytoplasmic male-sterile lines, male-sterile maintainer lines, and male-sterile restorer lines. In 1974, the first combination of hybrid rice was cultivated, followed by a success in seed production technology of hybrid rice in 1975, which marked that China became the first country to popularize and apply hybrid rice commercially on a large scale.

Based on the three-line theory in hybrid rice and the exchange and utilization of wild abortive cytoplasmic male sterile, Yan Long'an bred Zhenshan 97A in 1972; Xie Hua'an bred Minghui 63, a restorer line, and Shanyou 63, a strong heterotic combination successfully cultivated in 1980. Shanyou 63 had once been the hybrid rice variety the most widely planted in China for 15 consecutive years since 1986, with a maximum planting area of more than 100 million Chinese *mu* (1 *mu* is approximately equal to 666.7 m^2, similarly hereinafter) in a single year before. Subsequently, Zhu Yingguo and Zhou Kaida selected and bred red-lotus-type, Okan-type, and D-type hybrid rice, which is considered an approach to further enrich the Chinese hybrid rice breeding theory.

China began to popularize three-line hybrid rice on a large scale and achieved a huge increase in production in 1976, when 2.08 million *mu* of hybrid rice were planted nationwide, accounting for only 0.4% of the rice planting area. In 1977, it rose to 31 million *mu*, accounting for 5.8% of the rice planting area that year. In 1983, it exceeded 100 million *mu*, accounting for 20.3% of the rice planting area that year. In 1990, it exceeded 200 million *mu*, accounting for 49.8% of the rice planted area that year. Since 1991, the area covered with hybrid rice has stabilized at about 240 million *mu* accounting for more than 60% of the rice planting area.

With the research and implementation of high-yielding cultivation techniques, coupled with hybrid rice breeding, the single-cropping yield of hybrid rice has increased year by year. In 1976, the yield was 280 kg/*mu*, and in 1977 it reached 358.9 kg/*mu*. By 1983, it had surpassed 400 kg/*mu* and reached 440 kg/*mu* in 1988. Since then, it has remained stable at around 450 kg/*mu*, increasing by over 75 kg/*mu* compared with conventional rice.

According to the partial statistics from 1976 to 2020, China had promoted hybrid rice cultivation on a cumulative area of 610 million ha, resulting in an increase of 915 million tons of grain production. The annual increase in grain production owing to hybrid rice cultivation can feed an additional 80 million people, making significant contributions to food security in China.

Following the successful promotion and application of the three-line hybrid rice, the two-line hybrid rice has developed rapidly due to its greater potential for yield increase, better grain quality, and stronger resistance. Since its successful development in 1995, the planting area of two-line hybrid rice has continued to increase. In 1995, it was only 1.095 million *mu*, but by 2001, it had risen to 40 million *mu*, accounting for 17.2% of the hybrid rice planting area. By 2003, the planting area of two-line hybrid rice had increased to 49.5 million *mu*, with more than 80 photoperiod-thermo-sensitive genic male sterile (PTGMS) varieties and over 100 two-line hybrid combinations approved and applied in production. Generally speaking, the yield of two-line hybrid rice was 5%–10% higher than that of the current three-line hybrid rice. According to partial statistics, from 1993 to 2012, a total of 499.09 million *mu* of two-line hybrid rice had been promoted nationwide, resulting in an increase of 11.099 billion kg of rice. Among them, from 2010 to 2012, the promoted area reached 159.26 million *mu*, resulting in an increase of 3.5907 billion kg of rice.

II. The Super Hybrid Rice Development Course and Its Status Quo

In 1981, Japan first proposed a plan to implement rice breeding for ultra-high yield in the world, with a goal to spend 15 years breeding ultra-high-yielding varieties with a 50% yield increase than the popularized varieties at that time by 1995. However, due to the high difficulty and inappropriate technical approach, it has not been realized so far. In the early 1990s, the International Rice Research Institute launched the Breeding of New Plant-Type Rice Project, aiming to breed super rice with a potential yield of 12 t/ha by 2005. The proposal and implementation of this project received a strong international response, which was reported internationally under the title "Super Rice" by an American journalist. As a result, the breeding for ultra-high-yielding rice was known as super rice breeding. Subsequently, major rice-producing countries around the world began proposing and implementing their own super rice programs.

In 1996, a two-stage development plan for Chinese super hybrid rice was launched by the Chinese Ministry of Agriculture and Rural Affairs with goals determined. Taking the example of mid-season rice in the Yangtze River Basin, the first phase aimed to develop ultra-high-yielding rice varieties (combinations) that could achieve a continuous yield of 700 kg/*mu* in a hundred-*mu* demonstration plot in the same ecological zone for two consecutive years by 2000. These varieties should also possess resistance to two major sorts of disease and pest damage, with the main rice quality indicators meeting the national grade 2 high-quality rice standards. In the second phase, by 2005, the goal was to develop super rice varieties that could achieve a yield of 800 kg/*mu* in a large area with resistance to more than two sorts of disease and pest damage. Furthermore, the main rice quality indicators should meet the national grade 1 high-quality rice standards. To implement this plan, relevant departments in the country had organized collaboration among multiple agricultural research institutions nationwide. Currently, China has achieved tremendous success and is at the forefront of super rice breeding internationally. Taking the example of single-cropping rice, China has achieved the breeding goals of the first phase (700 kg/*mu*) of super rice research in 2000, the second phase (800 kg/*mu*) in 2004, and the third phase (900 kg/*mu*) in 2011.

Super rice can be divided into two categories: super conventional rice varieties and super hybrid rice combinations. It can also be classified into two types based on their advantages in yield, quality, and fertilizer tolerance: fertilizer-tolerant super rice with exceptionally high yields, good quality, and strong fertilizer tolerance; widely-adapted super hybrid rice with high yield, excellent quality, and strong adaptability.

The Super Rice Breeding Project has been initiated since the Ninth Five-Year Plan in China. With the assistance of numerous rice researchers, a batch of super rice varieties (combinations) confirmed by the Ministry of Agriculture and Rural Affairs had been successfully developed by 2020.

Among them are 48 indica three-line hybrid rice varieties, such as Guodao No. 1, Fengyuanyou 299, D You 527, Tianyou 998, Luoyou 8, etc.; 42 indica two-line hybrid rice varieties, such as Liangyoupei 9, Zhunliangyou 527, Y You 1, Zhuliangyou 819, Yangliangyou 6, etc.; and 1 three-line hybrid rice variety of japonica type, such as Liaoyou 1052.

Additionally, according to the regulations set by the Ministry of Agriculture and Rural Affairs for the identification of super rice varieties, some varieties have been withdrawn from the designation of super rice due to their failure to meet the required promotion area (some due to prolonged promotion time resulting in reduced area), including varieties such as Fengyou 299, Liangyoupei 9, Liaoyou 5218, III You 98, etc.

To fully unleash the yield-increasing potential of super hybrid rice, Academician Yuan Longping advocated in 2006 the strategic conception of the Planting Three to Produce Four Yield-Increasing Program of Super Hybrid Rice. By utilizing the achievements of existing super hybrid rice technology, the strategy aims to produce four units of grain from three units of farmland area, significantly increasing both the per-unit and total existing rice production, enhancing the economic benefits for farmers engaged in grain cultivation, and ensuring national food security. This strategic conception by Academician Yuan Longping gained tremendous support from the Hunan Provincial Government. From 2007 to 2016, the Planting Three to Produce Four Program was implemented as a major special project in Hunan Province for ten consecutive years. The implementation of the project has achieved both significant economic and social benefits, making great contributions to the sustainable and stable growth of grain production, the development of modern agriculture, and the construction of new rural areas in Hunan Province.

Based on the major achievements of super hybrid rice high-yielding demonstration and the yield gap of over 30% in rice yield per unit area in practice, Academician Yuan Longping had proposed the "Three-One" Grain High-Yielding Technology Engineering in 2014, which aimed to achieve an annual grain production of 1,200 kg/*mu* (approximately 0.067 ha) in high-yielding areas of southern China, with the goal of "feeding one person with three Chinese *fen* of land" (based on an annual production of 1,200 kg/*mu*, which corresponds to 360 kg of grain every three *fen* of land. This aligns with the national food security target of 360 kg of grain per person per year). The "Three-One" Grain High-Yielding Technology Engineering was officially established as a major special project of Hunan in 2017. By implementing the project, Hunan has been advanced to achieve an increase in total grain output and farmers' income.

III. The Development Strategy of Hybrid Rice

China is the first country in the world to successfully apply heterosis to rice, and the breeding technology of hybrid rice is also at the forefront internationally. However, everything in the world is always developing endlessly, following a spiral-shaped upward trajectory.

In 1987, Academician Yuan Longping proposed the Development Strategy of Hybrid Rice, which can be divided into three stages in terms of breeding methods: three-line method, two-line method, and one-line method. The development direction is from complexity to simplicity, with increasing efficiency. In terms of heterosis level, the breeding can be divided into three stages: inter-varietal hybrid, subspecies, and utilization of distant hybrid vigor. For the breeding of hybrid rice, there are three strategic development stages based on breeding methods and heterosis level; and each entry into a new stage represents a new breakthrough, pushing rice yield to a higher level.

1. The development of breeding methods for hybrid rice

1) *Three-line method.* It refers to the matching of three lines including the male sterile lines, maintainer lines, and restorer lines. The three-line method is a classic approach for breeding new combinations of hybrid rice. However, it also faces major problems such as complex breeding procedures and production processes, and long breeding cycles, which mainly manifested in the following aspects: First, the process of seed production is cumbersome. Second, the probability of breeding high-yielding, high-quality, and strong-resistant hybrid

rice combinations is low due to the limitations imposed by the restorer and maintainer relationships. Third, the yield remains stagnant. Fourthly, there is a lack of high-yielding, high-quality, and early-maturing hybrid long-grain indica rice variety and high-yielding, high-quality hybrid japonica rice combinations.

2) *Two-line method.* The two-line method involves hybrid rice with only two parents, which can be broadly classified into two types: PTGMS method and the chemical hybridization agents method. The former is which now commonly referred to. The discovery of rice with photoperiod-sensitive male sterile materials in 1973 and thermo-sensitive male sterile materials in 1987 marked another breakthrough in rice breeding, propelling the development of hybrid rice into a new stage. These materials can be broadly classified into two categories: One is the photoperiod-sensitive male sterile material represented by Nongken 58s, which is mainly controlled by a pair of recessive nuclear genes and is independent of cytoplasmic. For this type of material, it manifests male sterility under long-day high-temperature conditions and male fertility under short-day low-temperature conditions. The other category is the thermo-sensitive male sterile material represented by Annong s-1, which manifests male sterility under high temperatures and male fertility under relatively lower temperatures.

   The utilization of heterosis in the two-line method only requires two breeding materials, namely PGTMS lines and restorer lines, to produce hybrid seeds. For some reason, the PTGMS line can serve a dual purpose, eliminating the need for an additional seed production step compared to the three-line method. Furthermore, the PTGMS materials are controlled by recessive nuclear genes with simple genetic behavior. In theory, any excellent breeding material can be possibly developed into a PTGMS line. Moreover, more than 98% of the breeding materials in rice germplasm resources can be used as a restorer line in the two-line method, which greatly enhances the flexibility in selecting hybrid combinations and significantly increases the probability of breeding excellent combinations. Currently, a number of practical PTGMS lines have been successfully bred, such as Peiai 64s and Y58s. A series of mid-season and late-season combinations in the two-line method has also been successfully developed, such as Liangyoupei 9, Zhunliangyou 527, and Y Liangyou 1. Meanwhile, the propagation techniques of PTGMS lines and the seed production technology of hybrid combinations in the two-line method have been fully mastered.

3) *Single-line method.* The single-line method aims to breed non-segregating F_1 hybrids, which is a long-term strategic goal to fix the hybrid vigor and exemption from seed production.

2. The improvement of advantages of hybrid rice

Based on heterosis, the breeding of hybrid rice can be divided into three stages: intervarietal hybrid, subspecies, and utilization of distant hybridization.

1) *The heterosis of intervarietal hybrid.* Currently, the hybrid rice varieties used in production mainly belong to this category. In the 1970s, China made significant breakthroughs after the breeding of dwarf rice varieties by utilizing this type of hybrid vigor, which resulted in

a widespread yield increase of over 20%. However, due to the closer genetic relationship between varieties, there are great limitations in utilizing heterosis.

2) *The heterosis of subspecies.* Indica rice, japonica rice, and javanica rice are three subspecies of commonly cultivated rice. Due to the significant genetic differences between indica and japonica rice, the hybrid offspring between them show great potential in yielding. It has been a long-desired wish of breeders for years to utilize the strong heterosis between indica and japonica rice directly, but was difficult and challenging, mainly due to the low grain-setting rate of hybrids. However, the research conducted by Japanese scientist Ikehashi Hiroshi and others revealed the nature of the incompatibility of indica-japonica rice resulting in the low seed-setting rate of hybrids. Nowadays, the methods and materials to overcome this challenge have been basically identified and mastered, leading to the successful utilization of the heterosis between indica and japonica.
3) *The distant heterosis.* Distant hybridization can break the boundaries between rice varieties to some extent and facilitate the exchange of different genes. As a breeding technique, it is mainly used to introduce useful genes from different species to improve existing varieties. While the exploitation of distant heterosis presents greater challenges compared to other stages, advancements in biotechnology make it increasingly feasible. Among the most promising approaches for breeding distant hybrid rice are the utilization of apomixis and genetic engineering. The utilization of apomixis and genetic engineering may be the most promising approach and method for breeding distant hybrid rice.

There is a certain intrinsic relationship between the three breeding methods and the three levels of heterosis mentioned above. The three-line method is mainly suitable for breeding intervarietal hybrid combinations. Although it can also be used for combinations of subspecies, it is harder to achieve the desired results. The two-line method is applicable for breeding combinations of both intervarietal hybrid and subspecies, but it can better leverage its advantages when used for hybridization of subspecies. As for the utilization of distant heterosis, both the three-line method and the two-line method can utilize several favorable distant genes, but better results may only be achieved through the one-line method.

Everything is in eternal development. In 2018, Academician Yuan Longping further proposed the Five-Generation Development Strategy for Hybrid Rice. Hybrid rice has experienced rapid development from the first generation utilizing cytoplasmic male sterility line as a genetic tool in the three-line method to the second generation utilizing PTGMS as a genetic tool in the two-line method. Currently, research is underway to overcome the challenges of breeding the third generation of hybrid rice utilizing spontaneous genic male sterility as the genetic tool. Meanwhile, he believes that the strategy for hybrid rice development will continue to forge ahead toward the fourth generation involving C_4 hybrid rice and the fifth generation ensuring the heterosis of rice through utilizing apomixis.

IV. Practice and Case Analysis of Ultra-High Yield of Super Hybrid Rice

Since the initiation of the Chinese Super Rice Research in China in 1996, research on super hybrid rice has made rapid progress and achieved significant breakthroughs. In 1997, the pioneer combination of super hybrid rice, Peiai 64s/E32, jointly bred by the National Hybrid Rice Engineering Technology Research Center and Jiangsu Academy of Agricultural Sciences, was trial planted on a small scale of 3.6 *mu* in three locations in Jiangsu, with an average yield of 884 kg/*mu*. In 2002, Liangyou 293, the super rice combination bred by the National Hybrid Rice Engineering Technology Research Center, was planted in Longshan County on a demonstration basis of 127 *mu*, with an average yield of 817.3 kg/*mu*. In 2004, the semi-superior combination, Zhunlangyou 527, was planted in Guidong County and Rucheng County on a demonstration basis of 100 *mu*, with average yields of 842.1 kg/*mu* and 809.2 kg/*mu*, respectively. In 2005, the super rice seedling combination T98A/RB207-1 developed by the National Hybrid Rice Engineering Technology Research Center, was cultivated as amid-season rice in Jinshiqiao Town, Longhui County, Hunan Province. Through the acceptance and yield assessment by the Agricultural Commission of Hunan Province, the small-scale yield reached 902.2 kg/*mu*. On September 18, 2011, Y Liangyou 2, bred by the National Hybrid Rice Engineering Technology Research Center covered an area of 107.9 *mu* on a demonstration area of 100 *mu* in Leifeng Village, Yanggu'ao Township, Longhui County, Hunan Province. After yield assessment and acceptance by experts organized by the Ministry of Agriculture and Rural Affairs, the average yield reached 926.6 kg/*mu*, achieving the third-phase breeding goal of the Chinese Super Rice Program in advance.

With the continuous progress of the super rice project, a series of typical examples of ultra-high yield continued to emerge. For instance, the II Youhang 1, bred by the Fujian Academy of Agricultural Sciences, achieved a high yield of 928.3 kg/*mu* in Youxi County. Xieyou 9308, bred by the China National Rice Research Institute, achieved a yield of 818.8 kg/*mu*, setting the highest record for rice yield in Zhejiang Province. In 2014, the Y Liangyou 900 in Xupu County, Hunan Province, achieved an average yield of 1,026.7 kg/*mu* on a contiguous area of 100 *mu*, as shown in TABLE 1.1.

TABLE 1.1 Typical examples of super rice with ultra-high yield

The varieties/ combinations of super hybrid rice	Site	Acceptance institution	Measured yield (kg/*mu*)
Liangyou 293	Hunan Longhui Hundred-*mu* Demonstration Area	Agriculture Commission of Hunan Province	809.9
Zhunliangyou 527	Hunan Guidong Hundred-*mu* Demonstration Area	Agriculture Commission of Hunan Province	842.1

The varieties/ combinations of super hybrid rice	**Site**	**Acceptance institution**	**Measured yield (kg/*mu*)**
Zhunliangyou 527	Hunan Rucheng Hundred-*mu* Demonstration Area	Agriculture Commission of Hunan Province	809.2
Xieyou 9308	Zhejiang Yongchang Hundred-*mu* Demonstration Area	Ministry of Agriculture and Rural Affairs of China	818.8
Y Liangyou No. 1	Hunan Longhui Hundred-*mu* Demonstration Area	Ministry of Agriculture and Rural Affairs of China	926.6
II You 7954	Zhejiang Kaihua Hundred-*mu* Demonstration Area	Department of Science and Technology of Zhejiang Province	882
Zhongzheyou No. 1	Zhejiang Kaihua Hundred-*mu* Demonstration Area	Department of Science and Technology of Zhejiang Province	816
Guodao No. 6	Zhejiang Fuyang Hundred-*mu* Demonstration Area	Department of Science and Technology of Zhejiang Province	842
II Youming 86	Fujian Youxi Hundred-*mu* Demonstration Area	Agriculture Commission of Fujian Province	847.4
II Youhang No. 1	Fujian Shaxian Hundred-*mu* Demonstration Area	Agriculture Commission of Fujian Province	883.07
Y Liangyou 900	Hunan Xupu Hundred-*mu* Demonstration Area	Agriculture Commission of Hunan Province	1026.7
Xiangliangyou 900	Hebei Handan Hundred-*mu* Demonstration Area	Hebei Provincial Department of Science and Technology	1149.0
II Youming 86	Yunan Yongsheng Hundred-*mu* Demonstration Area	Fujian Academy of Sciences	1196.5
Xiangliangyou 900	Henan Guangshan Thousand-*mu* Demonstration Area	Agricultural Commission of Henan Province	913.9

CHAPTER 2

Characteristics of Super Hybrid Rice Varieties

I. Types of Super Hybrid Rice Varieties

Since the approval of the Research on Breeding and Cultivation Technology System of Super Rice in China scientific program by the Ministry of Agriculture and Rural Affairs and the Ministry of Science and Technology in 1996, significant progress has been made in the breeding of super rice varieties in China. A number of ultra-yielding rice varieties (combinations) with high-yielding, superior quality, and good resistance suitable for different ecological regions in China have been successfully cultivated. The National Agro-Tech Extension and Service Center has organized experts to establish standards for the confirmation of super rice varieties, and annual evaluations are conducted for rice varieties that meet the criteria. From 2005 to 2020, a total of 197 confirmed varieties of super hybrid rice were recommended to our country through experts' evaluations and official identification by the General Office of the Ministry of Agriculture and Rural Affairs. Among them, 64 varieties had been canceled the title of "Super Rice" due to the early approval year causing degradation or failure to meet the standard of super rice in terms of promotion area (TABLE 2.1).

TABLE 2.1 List of super rice varieties recognized by the Ministry of Agriculture and Rural Affairs

Year	Name ("*" refers to varieties whose title of "Super Rice" has been canceled)
2005	Xieyou 9308*, Guo 1*, Guo 3*, Zhongzheyou 1, Fengyou 299*, Jingyou 299*, Ⅱ Youming 86, Ⅱ Youhang No. 1*, Teyouhang No. 1*, D You 527*, Xieyou 527*, Ⅱ You 162*, Ⅱ You 7*, Ⅱ You 602, Tianyou 998, Ⅱ You 084*, Ⅱ You 7954*, Liangyoupei 9*, Zhunliangyou 527*, Liaoyou 5218*, Liaoyou 1052*, Ⅲ You 98*, Shengtai 1*, Shennong 265*, Shennong 606*, Shennong 016*, Jijing 88, Jijing 83*
2006	Songjing 9*, Tiejing 7*, Longjing 14*, Longdao 5*, Jijing 102*, Kendao 11*, Yongyou 6*, Guinongzhan, Zhongzao 22*, Wujing 15*, Liangyou 287, Zhuliangyou 819, Y Liangyou No. 1, Peiza Taifeng*, Xinliangyou No. 6, Qiannanyou 2058*, Yifeng 8*, Q You No. 6, Tianyou 122*, Jinyou 527*, D You 202*
2007	Ningjing 1*, Huaidao 9*, Qianchonglang 2*, Liaoxing 1, Chujing 27, Longjing 18*, Yuxiangyouzhan, Xinliangyou 6380, Fengliangyou No. 4 (Wandao 187), Nei 2 You 6*, Ganxin688*, Ⅱ Youhang 2*
2009	Longjing 21, Huaidao 11*, Zhongjiazao 32*, Yangliangyou No. 6, Luliangyou 819, Fengliangyouxiang No. 1, Luoyou No. 8, Rongyou 3*, Jinyou 458*, Chunguang 1*
2010	Xindao 18*, Yangjing 4038*, Ningjing 3*, Nanjing 44*, Zhongjiazao 17, Hemeizhan, Guiliangyou 2, Peiliangyou 3076*, Wuyou 308, Wufengyou T025, Xinfengyou 22*, Tianyou 3301
2011	Shennong 9816, Wuyunjing 24*, Nanjing 45*, Yongyou 12, Lingliangyou 268, Zhunliangyou 1141*, Huiliangyou No. 6, 03 You 66*, Teyou 582
2012	Chujing 28*, Lianjing 7, Zhongzao 35, Jingnongsimiao, Zhunliangyou 608, Shengliangyou 5814, Guangliangyouxiang 66, Jingyou 785*, Dexiang 4103, Q you 8*, Tianyou Huazhan, Yiyou 673, Shengyou 9516
2013	Longjing 31, Songjing 15, Zhengtao 11, Yangjing 4227, Ningjing 4, Zhongzao 39, Y Liangyou 087, Tianyou 3618, Tianyou Huazhan, Zhong 9 You 8012, H You 518, Yongyou 15
2014	Longjing 39, Liantao 1, Changbai 25, Nanjing 5055, Nanjing 49*, Wuyunjing 27, Y Liangyou No. 2, Y Liangyou 5867, Liangyou 038, C Liangyou Huazhan, Guangliangyou 272, Liangyou No. 6, Liangyou 616, Wufengyou 615, Shengtaiyou 722, Nei 5 You 8015, Rongyou 225, F You 498

Year	Name ("*" refers to varieties whose title of "Super Rice" has been canceled)
2015	Yangyujing 2, Nanjing 9108, Zhengdao 18, Huahang 31, H Liangyou 991, N Liangyou 2, Yixiangyou 2115, Shengyou 1029, Yongyou 538, Chunyou 84, Zheyou 18
2016	Jijing 511, Nanjing 52, Huiliangyou 996, Shenliangyou 870, Deyou 4727, Fengtianyou 553, Wuyou 662, Jiyou 225, Wufengyou 286, Wuyouhang 1573
2017	Nanjing 0212, Chujing 37, Y Liangyou 900, Longliangyou Huazhan, Shengliangyou 8386, Y Liangyou 1173, Yixiang 4245, Jifengyou 1002, Wuyou 116, Yongyou 2640
2018	Longliangyou 1988, Shengliangyou 136, Jingliangyou Huazhan Wuyou 369, Neixiang 6 You 9, Shuyou 217, Luyou 727, Jiyou 615, Wuyou 1179, Yongyou 1540
2019	Ningjing 7, Shenliangyou 862, Longliangyou 1308, Longliangyou 1377, Heliangyou 713, Y Liangyou 957, Longliangyou 1212, Jinliangyou 1212, Huazheyou No. 1, Wantaiyou 3185
2020	Suken 118, Zhongzu 143, Yongyou 7850, Jinliangyou 1988, Jiafengyou 2, Huazheyou 71, Funongyou 676, Jiyouhang 1573, Taiyou 871, Longfengyou 826, Jingyou Huazhen

Note: According to the Measures for Identifying Super Rice Varieties (Agricultural Office [2008], No. 38), the qualification of some varieties for the title of Super Rice has been canceled due to the failure to meet the requirements of promotion area, marked with "*" in the table.

According to the types of super rice varieties, they can be divided into conventional japonica rice, conventional indica rice, three-line indica super hybrid rice, three-line japonica super hybrid rice, two-line indica super hybrid rice, and super hybrid rice of indica-japonica hybridization.

1. The three-line indica super hybrid rice

Varieties mainly include Jiafengyou 2, Huazheyou 71, Funongyou 676, Jiyouhang 1573, Taiyou 871, Longfengyou 826, Jingyou Huazhen, Huazheyou No. 1, Wantaiyou 3158, Wuyou 369, Neixiang 6 You 9, Shuyou 217, Huyou 727, Jiyou 615, Wuyou 1179, Yixiang 4245, Jifengyou 1002, Wuyou 116, Deyou 4727, Fengtianyou 553, Wuyou 662, Jiyou 225, Wufengyou 286, Wuyouhang 1573, Yixiangyou 2115, Shenyou 1029, F You 498, Rongyou 225, Nei 5 You 8015, Shengtaiyou 722, Wufengyou 615, Tianyou 3618, Tianyou Huazhan, Zhong 9 You 8012, H You 518, Dexiang 4103, Yiyou 673, Shenyou 9516,

Teyou 582, Wuyou 308, Wufengyou T025, Tianyou 3301, Luoyou No. 8, Q-you No. 6, Zhongzheyou 1, II Youming 86, II You 602, and Tianyou 998.

2. The three-line japonica super hybrid rice

Currently, there is only one recognized variety of three-line japonica super hybrid rice, named Liaoyou 1052.

3. The two-line indica super hybrid rice

Varieties mainly include Jinliangyou 1988, Shenliangyou 862, Longliangyou 1308, Longliangyou 1377, Heliangyou 713,Y Liangyou 957, Longliangyou 1212, Jinliangyou 1212, Longliangyou 1988, Shenliangyou 136, Jinliangyou Huazhan, Y Liangyou 900, Longliangyou Huazhan, Shenliangyou 8386, Y Liangyou 1173, Weiliangyou 996, Shenliangyou 870, H Liangyou 991, N Liangyou 2, Liangyou 616, Liangyou No. 6, Guangliangyou 272, C Liangyou Huazhan, Liangyou 038, Y Liangyou 5867, Y Liangyou No. 2, Y Liangyou 087, Zhunliangyou 608, Shenliangyou 5814, Guangliangyouxiang 66, Lingliangyou 268, Huiliangyou No. 6, Guiliangyou 2, Yangliangyou No. 6, Luliangyou 819, Fengliangyouxiang No. 1, Xinliangyou 6380, Fengliangyou No. 4, Y You 1, Zhuliangyou 819, Liangyou 287, and Xinliangyou No. 6.

4. The super hybrid rice of indica-japonica hybridization

Varieties mainly include Yongyou 7850, Yongyou 1540, Yongyou 2640, Yongyou 538, Chunyou 84, Zheyou 18, Yongyou 15, and Yongyou 12.

II. Biological Characteristics of Super Hybrid Rice Varieties

There are 100 varieties selected from the list of rice varieties that have been recognized as super hybrid rice since 2005, including those whose Super Rice title weas canceled. The main characteristics of these varieties are briefly introduced as follows.

1. Liangyoupei 9

The two-line indica hybrid rice bred by the Crop Research Institute of Jiangsu Academy of Agricultural Sciences and the Hybrid Rice Research Center in Hunan was recognized as a super rice variety in 2005, but the designation was revoked in 2020. It has an average growth period of 150 days in the southern rice-growing regions and is susceptible to bacterial leaf blight and rice blast diseases. The rice quality characteristics include a head rice rate of 53.6%, a chalky rice rate of 35%, a chalkiness degree of 4.3%, a gel consistency of 68.8 mm, and an amylose content of 21.2%. The rice has excellent eating quality. In field trials conducted in the southern rice-growing regions, the average yield ranged from 525.8 to 576.9 kg/*mu*, while in Jiangsu Province, the average yield was 625.5 kg/*mu*. It has broad adaptability and is suitable for cultivation as a single-cropping rice variety in Guizhou, Yunnan, Sichuan, Chongqing, Hunan, Hubei, Jiangxi, Anhui, Jiangsu, Zhejiang, Shanghai, and Henan. From 2005 to 2015, the cumulative promotion area of this variety reached 43.45 million *mu* (approximately 2.9 million ha).

2. Guodao No. 1

The three-line indica hybrid rice bred by the China National Rice Research Institute was identified as a super rice variety in 2005 and had its title removed in 2019. It is suitable for cultivation as a double-cropping late-season rice variety in the middle and lower reaches of the Yangtze River with an average growth period of 120.6 days. It has good resistance to rice blast disease and bacterial leaf blight, and the quality of the rice is excellent. The head rice rate is 55.9%, the length-to-width ratio is 3.4, the chalky rice rate is 21%, the chalkiness degree is 3.4%, the gel consistency is 64 mm, and the amylose content is 21.2%. During the two-year regional trials from 2002 to 2003, the average yield was 458.22 kg/*mu*, and during the production trial in 2003, the average yield was 433.60 kg/*mu*. It is suitable for cultivation as a double-cropping late-season rice variety in the central and northern parts of Guangxi, the central and northern parts of Fujian, the central and southern parts of Jiangxi, the central and southern parts of Hunan, and the southern part of Zhejiang. This variety has been promoted on a cumulative area of 7.04 million *mu* from 2005 to 2015.

3. Fengyou 299

The three-line indica hybrid rice variety, bred by Hunan Hybrid Rice Research Center, was recognized as a super rice variety in 2005, but had its title revoked in 2020. It is suitable for cultivation as a double-cropping late-season rice variety in Hunan Province, with an average growth period of 114 days. It is moderately resistant to rice blast disease and bacterial leaf blight and has medium cold tolerance. The rough rice rate is 83.1%, the milled rice rate is 75.6%, and the head rice rate is 66.9%. The length-to-width ratio is 3.0, and the chalky rice rate is 23%, with a degree of 2.6%. The average yield per *mu* was 471.6 kg during the two-year regional trial from 2002 to 2003, and it is suitable for cultivation as a double-cropping late-season rice variety in Hunan. From 2005 to 2015, the cumulative promotion area of this variety reached 13.59 million *mu*.

4. Teyouhang No. 1

The three-line indica hybrid rice developed by the Rice Research Institute of Fujian Academy of Agricultural Sciences was recognized as a super rice variety in 2005. However, its title was revoked in 2018. It is cultivated in the upper reaches of the Yangtze River as a single-cropping mid-season rice variety. The average growth period is 150.5 days, and it is susceptible to rice blast and bacterial leaf blight diseases. The quality of the rice is average, with a head rice rate of 63.5%, a length-to-width ratio of 2.4, a chalky rice rate of 83%, a chalkiness degree of 16.2%, a gel consistency of 62 mm, and an amylose content of 20.7%. During the regional trials from 2002 to 2003, the average yield was 591.72 kg/*mu*. In the production trials of 2004, the average yield was 573.28 kg/*mu*. It is suitable for cultivation in the low- and mid-altitude rice areas of Yunnan, Guizhou, Chongqing (excluding the Wuling Mountain area), Sichuan Pingba hilly rice area, and the southern part of Shaanxi. From 2005 to 2015, this variety has been promoted on an accumulated area of 4.58 million *mu* (approximately 305,333 ha).

5. D You 527

D You 527 is a three-line indica hybrid rice variety bred by the Rice Research Institute of Sichuan Agricultural University. It was recognized as a super hybrid rice variety in 2005 but had its title revoked in 2018. Its average growth period is 153.1 days in the upper reaches of the Yangtze River and 143.5 days in the middle and lower reaches. It has above-average rice quality, with resistance to rice blast disease but not to bacterial leaf blight and brown planthopper. The head rice rate is 52.1%, and the length-to-width ratio is 3.2, with a 43.5% chalky rice rate and a 7.0% chalkiness degree. Besides, the gel consistency is 51 mm with 22.7% amylose content. During the regional trials in the mid-season late-maturing group in the Yangtze River Basin from 2000 to 2001, it achieved an average yield of 591.36 kg/*mu*. In the production trials of 2001, the average yield was 648.31 kg/*mu* in the upper reaches of the Yangtze River and 567.7 kg/*mu* in the middle and lower reaches. It is suitable for mid-season rice cultivation in the Yangtze River Basin (excluding the Wuling Mountain area), Sichuan, Chongqing, Hubei, Hunan, Zhejiang, Jiangxi, Anhui, Shanghai, and Jiangsu. It can also be planted in Yunnan and Guizhou provinces below a 1,100 m altitude and in Xinyang, Henan Province, and Hanzhong, Shaanxi Province. From 2005 to 2015, this variety has been promoted on a cumulative area of 10.33 million *mu* (approximately 688,667 ha).

6. Liangyou 287

The two-line indica hybrid rice bred by the School of Life Sciences of Hubei University was recognized as a super rice variety in 2006. It participated in the regional trial of early rice varieties in Hubei Province from 2003 to 2004. The rice quality was determined by the Food Quality Monitoring Center of the Ministry of Agriculture China. The results showed that it had a brown rice rate of 80.4%, a head rice rate of 65.3%, a chalky rice rate of 10%, a chalkiness degree of 1.0%, an amylose content of 19.5%, a gel consistency of 61 mm, a length-to-width ratio of 3.5, and its main physical and chemical indicators met the national standards of grade 1 high-quality rice according to *High Quality Paddy*. During the two-year regional trial, it achieved an average yield of 458.27 kg/*mu* and had a full growth period of 113.0 days. It was identified as highly susceptible to panicle and stem blast diseases and susceptible to white leaf spot disease, making it suitable for early-season rice cultivation in Hubei Province.

7. II Youming 86

II Youming 86, a three-line indica hybrid rice bred by the Sanming Agricultural Science Research Institute, was recognized as a super rice variety in 2005. It has an average growth period of 150.8 days for mid-season rice planting and 128–135 days for double-cropping late rice. The variety has a head rice rate of 56.2%, a chalky rice rate of 78.8%, a chalkiness degree of 18.9%, a gel consistency of 46 mm, and an amylose content of 22.5%. It is susceptible to rice blast and bacterial leaf blight diseases, with a disease rating of 4.5 and 8, respectively, and a level 7 infestation rating for rice planthoppers. In 1999, it participated in the regional trials for the mid-season late-maturing group of indica rice in the national southern rice area, with an average yield of 632.18 kg/*mu*. In the production trials of 2000, it achieved an average yield of 565.4 kg/*mu*, and in 2001, it averaged 581.2 kg/*mu*. This variety exhibits late maturity, high yield, and broad adaptability, making it suitable for single-cropping mid-season rice cultivation in the Yangtze River Basin of Guizhou, Yunnan, Sichuan, Chongqing, Hunan, Hubei, Zhejiang, Shanghai, Anhui, Jiangsu, and in the southern part of Henan Province and the Hanzhong region of Shanxi.

8. Tianyou 998

Tianyou 998, a three-line indica hybrid rice bred by the Rice Research Institute of Guangdong Academy of Agricultural Sciences, was recognized as a super rice variety in 2005. It is suitable for double-cropping late-season rice cultivation in the middle and lower reaches of the Yangtze River, with an average growth period of 117.7 days. It exhibits an average resistance level of 3.3 against rice blast disease, with a maximum level of 9 and a resistance frequency of 90%. It has a level 7 resistance against bacterial leaf blight. The head rice rate is 56.7%, with a length-to-width ratio of 3.1. The chalky rice rate is 27% with a 2.5% chalkiness degree, and the gel consistency is 59 mm with 23.0% amylose content, meeting the national standards of grade 3 high-quality rice. In the regional trials for late-season indica rice in the middle and lower reaches of the Yangtze River in 2004 and 2005, it achieved an average yield of 512.62 kg/*mu*. In thc production trials of 2005, the average yield was 478.44 kg/*mu*. This variety is suitable for both early- and late-season rice planting in South China and for post-season rice planting in the Yangtze River Basin.

9. Liaoyou 1052

The three-line japonica hybrid rice, bred by the Rice Research Institute of Liaoning Academy of Agricultural Sciences, was recognized as a super rice variety in 2005. It has a growth period of approximately 160 days and belongs to the mid-late maturing type. The brown rice rate is 81.7%, the milled rice rate is 73.2%, and the head rice rate is 65.1%. The grain length is 5.3 mm with a length-to-width ratio of 2.0. The chalky rice rate is 37%, and the chalkiness degree is 3.5%. It has a transparency level of grade 2, an alkali digestion value of grade 7, a gel consistency of 66 mm, an amylose content of 17.6%, and a protein content of 7.4%. The rice has a unique aroma. It exhibits resistance against stem blast disease. In the regional trials conducted in the province in 2001 and 2002, its average yield was 637.0 kg/*mu*. In the production trials in the province in 2003, the average yield was 678.3 kg/*mu*. This variety is suitable for cultivation in Shenyang, Liaoyang, Tieling, Kaiyuan, Anshan, Yingkou, Wafangdian, and other areas within Liaoning Province. It is also suitable for cultivation in Xinjiang, Ningxia, Hebei, Shaanxi, Shanxi, and other regions outside of Liaoning Province.

10. Zhuliangyou 819

The two-line indica hybrid rice bred by the Hunan Yahua Seed Industry Scientific Research Institute was recognized as a super rice variety in 2006. It has a complete growth period of approximately 106 days when cultivated as a double-cropping early-season rice in Hunan Province. Resistance identification in Hunan Province showed that it has a level 5 resistance against leaf blast, a level 5 against panicle blast, a susceptibility to rice blast disease, and a level 5 against bacterial leaf blight. The rice quality testing results are as follows: the brown rice rate is 81.8%, the milled rice rate is 72.2%, and the head rice rate is 68.0%. The grain length is 6.5 mm with a length-to-width ratio of 3.0. The chalky rice rate is 60%, and the chalkiness degree is 9.9%. It has a transparency level of grade 2, an alkali digestion value of grade 4.9, a gel consistency of 60 mm, an amylose content of 22.1%, and a protein content of 10.8%. In the regional trials conducted in Hunan Province in 2003 and 2004, the average yield per *mu* was 470.48 kg with a daily yield of 4.42 kg. This variety is suitable for double-cropping early-season rice cultivation in Jiangxi and Hunan provinces.

11. Y Liangyou No. 1

Y Liangyou No. 1, a two-line indica hybrid rice bred by Hunan Hybrid Rice Research Center, was recognized as a super rice variety in 2006. It is primarily cultivated as a single-cropping mid-season rice in the upper reaches of the Yangtze River, with an average growth period of 160.8 days. The comprehensive index of rice blast disease is 6.4, with the highest level of panicle blast disease reaching level 7. The brown planthopper infestation is at level 9, making it highly susceptible to both rice blast disease and brown planthopper. As for rice quality indicators, the milled rice rate is 67.2%, the length-to-width ratio is 2.8, the chalky rice rate is 29.0%, the chalkiness degree is 4.3%, the gel consistency is 80 mm, and the amylose content is 17.2%, meeting the national standards of grade 3 high-quality rice according to *High Quality Paddy*. In the regional trials for mid-season indica rice conducted in the upper reaches of the Yangtze River for the years 2010 and 2011, the average yield per *mu* was 582.5 kg. In the production trials conducted in 2012, the average yield per *mu* was 619.5 kg. This variety is suitable for single-cropping mid-season rice cultivation in the low- and mid-altitude indica rice regions of Yunnan and Chongqing (except for the Wuling Mountain area), the hilly rice regions of Pingba in Sichuan, and the southern part of Shaanxi. It is also suitable for early-season rice cultivation in the double-cropping rice regions of Hainan, southern Guangxi, central and southwestern Guangdong, southern Fujian, as well as for the single-cropping mid-season rice cultivation in the Yangtze River Basin of Jiangxi, Hunan, Hubei, Anhui, Zhejiang, Jiangsu (except for the Wuling Mountain area), and in the northern part of Fujian and the southern part of Henan Province. It is not suitable for cultivation in areas with a high incidence of rice blast disease.

12. Xinliangyou No. 6

The two-line indica hybrid rice bred by Anhui Quanyin Agricultural High-Tech Research Institute was recognized as a super rice variety in 2006. It is primarily cultivated as a single-cropping intermediate rice in the middle and lower reaches of the Yangtze River, with an average growth period of 130.1 days. The comprehensive index of rice blast disease is 6.6, with the highest level of panicle blast disease reaching level 9, and bacterial leaf blight reaching level 5. As for rice quality indicators, the head rice rate is 64.7%, the length-to-width ratio is 3.0, the chalky rice rate is 38%, the chalkiness degree is 4.3%, the gel consistency is 54 mm, and the amylose content is 16.2%. In the regional trials for mid-season late-maturing indica rice conducted in the middle and lower reaches of the Yangtze River for the years 2005 and 2006, the average yield per *mu* was 572.39 kg. In the production trials conducted in 2006, the average yield per *mu* was 549.71 kg. This variety is suitable for cultivation in the Yangtze River Basin of Jiangxi, Hunan, Hubei, Anhui, Zhejiang, and Jiangsu (except for the Wuling Mountain area), and in the northern part of Fujian and the southern part of Henan.

13. Q You No. 6

Q You No. 6, a three-line indica hybrid rice bred by Chongqing Seed Company, was recognized as a super rice variety in 2006. It is primarily cultivated as a single-cropping mid-season rice in the upper reaches of the Yangtze River, with an average growth period of 153.7 days. The resistance levels are as follows: an average of 6.4 for rice blast disease, with a maximum level of 9, and a resistance frequency of 75.0%. As for rice quality indicators, the head rice rate is 65.6%, the length-to-width ratio is 3.0, the chalky rice rate is 22%, the chalkiness degree is 3.6%, the gel consistency is 58 mm, and the amylose content is 15.2%, meeting the national standard of grade 3 for high-quality rice according to *High Quality Paddy*. It participated in the regional trials for mid-season late-maturing indica rice conducted upstream of the Yangtze River in 2004 and 2005 achieving an average yield per *mu* of 598.35 kg. In the production trials conducted in 2005, its average yield per *mu* was 556.66 kg. This variety is suitable for cultivation as single-cropping mid-season rice in the low- and mid-altitude indica rice regions of Yunnan, Guizhou, Hubei, Hunan, Chongqing (except for the Wuling Mountain area), the Pingba hilly rice region in Sichuan, and the southern rice region in Shaanxi.

14. Fengliangyou No. 4

Fengliangyou No. 4, a two-line indica hybrid rice bred by Hefei Fengle Seed Industry Co. Ltd., was recognized as a super rice variety in 2007. In the middle and lower reaches of the Yangtze River, the average growth period was 135.3 days. The resistance levels are as follows: a comprehensive index of 6.2 for rice blast disease, with a maximum loss rate of 9 for panicle blast disease, 7 for bacterial leaf blight, and 9 for brown planthopper. As for rice quality indicators, the head rice rate is 60.3%, the length-to-width ratio is 2.9, the chalky rice rate is 21%, the chalkiness degree is 2.9%, the gel consistency is 75 mm, and the amylose content is 16.1%, meeting the national standard of grade 2 high-quality rice according to *High Quality Paddy*. In the regional trials for mid-season late-maturing indica rice conducted in the middle and lower reaches of the Yangtze River for the years 2007 and 2008, the average yield per *mu* was 606.40 kg. In the production trials conducted in 2008, the average yield per *mu* was 575.19 kg. This variety is suitable for cultivation as single-cropping mid-season rice in the Yangtze River Basin rice regions (except for the Wuling Mountain area) of Jiangxi, Hunan, Hubei, Anhui, Zhejiang, and Jiangsu, and in the northern part of Fujian and the southern part of Henan.

15. Yangliangyou No. 6

The two-line indica hybrid rice variety developed by the Agricultural Science Research Institute in Lixiahe, Jiangsu was recognized as a super rice variety in 2009. It can be planted as single-cropping mid-season rice in the middle and lower reaches of the Yangtze River, with an average growth period of 134.1 days. Its resistance levels are as follows: the average index for rice blast disease is 4.8, with a maximum of 7; for bacterial leaf blight, it is 3; and for brown planthopper, it is 5. The main rice quality indicators are as follows: a head rice rate of 58.0%, a length-to-width ratio of 3.0, a chalky rice rate of 14%, a chalkiness degree of 1.9%, a gel consistency of 65mm, and an amylose content of 14.7%. In the two years 2002 and 2003, it participated in the regional trials for the mid-season late-maturing high-quality indica rice group A in the middle and lower reaches of the Yangtze River, with an average yield of 555.98 kg/*mu*. In 2004, during the production trial, its average yield was 555.72 kg/*mu*. This variety is suitable for cultivation as single-cropping mid-season rice in the Yangtze River Basin in Fujian, Jiangxi, Hunan, Hubei, Anhui, Zhejiang, and Jiangsu (except for the Wuling Mountains area) and in the southern rice regions of Henan.

16. Luliangyou 819

The two-line indica hybrid rice bred by Hunan Yahua Seed Industry Science Research Institute was recognized as a super rice variety in 2009. It can be planted as double-cropping early-season rice in the middle and lower reaches of the Yangtze River, with an average growth period of 107.2 days. The resistance levels are as follows: the comprehensive index for blast disease is 3.9, with a maximum loss rate of 7 for panicle blast and a resistance frequency of 55%. It also has a grade 7 resistance for bacterial leaf blight, grade 5 of brown planthopper, and grade 7 of white-backed planthopper. The main rice quality indicators are a head rice rate of 59.0%, a length-to-width ratio of 3.4, a chalky rice rate of 72%, a chalkiness degree of 8.1%, a gel consistency of 59 mm, and an amylose content of 20.4%. In the years 2006 and 2007, it participated in the regional trials for early- and mid-early-season indica rice varieties in the middle and lower reaches of the Yangtze River, with an average yield of 508.0 kg/*mu*. During the production trial in 2007, the average yield was 455.0 kg/*mu*. This variety is suitable for cultivation as an early-season rice variety in Jiangxi, Hunan, Hubei, Anhui, and Zhejiang.

17. Fengliangyouxiang No. 1

Fengliangyouxiang No. 1, a two-line indica hybrid rice bred by Hefei Fengle Seed Industry Co. Ltd., was recognized as a super rice variety in 2010. It is cultivated as a single-cropping mid-season rice variety in the middle and lower reaches of the Yangtze River Basin, with an average growth period of 130.2 days. Its comprehensive index for rice blast disease is 7.3, with a maximum rating of 9 for panicle blast damage. Its average rating for bacterial leaf blight is 6, with a maximum rating of 7. As for rice quality indicators, the milled rice rate is 61.9%, the length-to-width ratio is 3.0, the chalky grain rate is 36%, the chalkiness degree is 4.1%, the gel consistency is 58 mm, and the amylose content is 16.3%, meeting the national standards of grade 3 high-quality rice according to *High Quality Paddy*. In the regional trials conducted for mid-season late-maturing indica rice in the middle and lower reaches of the Yangtze River in 2004 and 2005, the average yield per *mu* was 568.70 kg. In the production trials conducted in 2006, the average yield per *mu* was 570.31 kg. This variety is suitable for cultivation as a single-cropping mid-season rice variety in the Yangtze River Basin rice regions (except for the Wuling Mountain area) of Jiangxi, Hunan, Hubei, Anhui, Zhejiang, Jiangsu, the northern part of Fujian, and the southern part of rice-growing region in Henan.

18. Luoyou No. 8

The three-line indica hybrid rice bred by Wuhan University was recognized as a super rice variety in 2009. It is planted as a single-cropping rice in the middle and lower reaches of the Yangtze River, with an average growth period of 138.8 days. Resistance: the comprehensive index of blast disease is 5.1, with the highest level of rice blast damage at level 9 and bacterial leaf blight disease at level 7. The main indicators of rice quality are a head rice rate of 61.4%, a length-to-width ratio of 3.1, a chalky rice rate of 22%, a chalkiness degree of 4.1%, a gel consistency of 65 mm, and an amylose content of 22.7%, meeting the national standard of grade 3 high-quality rice according to *High Quality Paddy*. It participated in the regional trials for mid-season late-maturing indica rice varieties in the middle and lower reaches of the Yangtze River in 2004 and 2005, with an average yield of 568.50 kg/*mu*. In 2005, it participated in production trials with an average yield of 538.19 kg/*mu*. This variety is suitable for cultivation as a single-cropping mid-season rice variety in the Yangtze River Basin (except for the Wuling Mountain area) in Jiangxi, Hunan, Hubei, Anhui, Zhejiang, Jiangsu, the northern part of Fujian, and the southern part of rice-growing areas in Henan.

19. Guiliangyou No. 2

The two-line indica hybrid rice bred by the Rice Research Institute of Guangxi Academy of Agricultural Sciences was recognized as a super rice variety in 2010. It is suitable for early rice cultivation in southern Guangxi with a total growth period of 124 days. The main quality indicators include a brown rice rate of 82.7%, a head rice rate of 69.0%, a length-width ratio of 2.7, a chalky rice rate of 86%, a chalkiness degree of 15.0%, a gel consistency of 80 mm, and an amylose content of 24.4%. It has a level 6 resistance against seedling blast, a level 7 against panicle blast, a panicle blast loss index of 46.2%, and a comprehensive rice blast resistance index of 6.8; it also has a level 6 resistance against bacterial leaf blight type IV with a level 5 against type V. It participated in the regional trials for early-season late-maturing varieties in the southern Guangxi rice production area in 2006 and 2007, with an average yield of 511.12 kg/*mu*. In 2007, it participated in production trials with an average yield of 559.38 kg/*mu*, suitable for cultivation as an early-season rice variety in the southern Guangxi rice production area.

20. Wuyou 308

The three-line indica hybrid rice bred by the Rice Research Institute of Guangdong Academy of Agricultural Sciences was recognized as a super rice variety in 2010. It is suitable for double-cropping late-season rice planting in the middle and lower reaches of the Yangtze River, with an average growth period of 112.2 days. Resistance: the comprehensive index of rice blast disease is 5.1, with the highest level of panicle blast loss rate at level 9, and a resistance frequency of 85%; the average level of bacterial leaf blight is 6, with the highest level at 7; the level of brown planthopper resistance is 5. The main indicators of rice quality include a head rice rate of 59.1%, a length-to-width ratio of 2.9, a chalky rice rate of 6%, a chalkiness degree of 0.8%, a gel consistency of 58 mm, and an amylose content of 20.6%, meeting the national standard of grade 1 high-quality rice according to *High Quality Paddy*. In the regional trials of late-season early-maturing indica rice varieties in the middle and lower reaches of the Yangtze River in 2006 and 2007, the average yield per *mu* was 504.5 kg. In the production trial in 2007, the average yield per *mu* was 511.7 kg, suitable for cultivation as a late-season rice variety in the rice-growing areas south of the Yangtze River in Jiangxi, Hunan, Zhejiang, Hubei, and Anhui.

21. Wufengyou T025

The three-line indica hybrid rice bred by the School of Agricultural Science of Jiangxi Agricultural University was recognized as a super rice variety in 2010. It is suitable for cultivation as a double-cropping late-season rice variety in the middle and lower reaches of the Yangtze River, with an average growth period of 112.3 days. It has a comprehensive resistance index of rice blast disease of 5.5, with the highest panicle blast loss rate reaching level 9; it also has a level 7 resistance against bacterial leaf blight and level 9 against brown planthopper. The main indicators of grain quality include a head rice rate of 56.1%, a length-width ratio of 2.9, a chalky rice rate of 29%, a chalkiness degree of 4.7%, a gel consistency of 52 mm, and an amylose content of 22.5%, reaching the national standard of grade 3 high-quality rice according to *High Quality Paddy*. In the regional trials of late-season early-maturing indica rice varieties in the middle and lower reaches of the Yangtze River in 2007 and 2008, its average yield per *mu* was 501.3 kg. In the production trial in 2009, the average yield per *mu* was 490.11 kg. This variety is suitable for cultivation as a late-season rice variety in the rice-growing areas south of the Yangtze River in Jiangxi, Hunan, Hubei, Zhejiang, and Anhui.

22. Tianyou 3301

The three-line indica hybrid rice bred by the Biotechnology Research Institute of Fujian Academy of Agricultural Sciences and the Rice Research Institute of Guangdong Academy of Agricultural Sciences was recognized as a super rice variety in 2010. It is cultivated as a single-cropping rice variety in the middle and lower reaches of the Yangtze River, with an average growth period of 133.3 days. It has a comprehensive resistant index of blast disease of 3.3, with the highest panicle blast loss rate reaching level 5; it also has a level 9 resistance against bacterial leaf blight, level 7 against brown planthopper, and general cold tolerance. The main indicators of rice quality include a head rice rate of 47.9%, a length-width ratio of 3.1, a chalky rice rate of 36%, a chalkiness degree of 6.0%, a gel consistency of 79 mm, and an amylose content of 23.2%. In the regional trials of the middle and lower reaches of the Yangtze River for the late-maturing mid-season indica rice varieties in 2007 and 2008, the average yield per *mu* was 598.3 kg. In the production trials in 2009, the average yield per *mu* was 581.1 kg. This variety is suitable for cultivation as single-cropping mid-season rice in the Yangtze River Basin (except for the Wuling Mountain area) in Jiangxi, Hunan, Hubei, Anhui, Zhejiang, Jiangsu, the northern part of Fujian, and the southern part of Henan rice-growing areas.

23. Yongyou 12

The three-line japonica hybrid rice bred by the Ningbo Academy of Agricultural Sciences was recognized as a super rice variety in 2011. It participated in the regional trials of single-cropping late-season japonica rice varieties in Zhejiang Province in 2007 and 2008, with an average yield of 565.4 kg/*mu*. In 2009, the average yield in the provincial production trials was 603.7 kg/*mu*. The average full growth period was 154.1 days over the two years. According to the resistance identification conducted by the Plant Microbiology Institute of the Provincial Academy of Agricultural Sciences in 2007 and 2008, the average leaf blast level was 2.2, the average panicle blast level was 3.1, and the panicle blast loss rate was 4.1%. The comprehensive indices were 1.9 and 3.2, respectively. The average resistance level against bacterial leaf blight was 3.5, and the average level against brown planthopper level 7.0. According to the rice quality testing conducted by the Rice and Product Quality Supervision and Testing Center of the Ministry of Agriculture and Rural Affairs in 2007 and 2008, the average head rice rate was 68.8%, the length-width ratio was 2.1, the chalky rice rate was 29.7%, the chalkiness degree was 5.1%, the transparency level was 3, the gel consistency was 75.0 mm, and the amylose content was 14.7%. The rice quality indicators over the two years reached the grade 4 and 5 respectively, as specified by the Department of Edible Rice Variety Quality of the Ministry of Agriculture and Rural Affairs. This variety is suitable for cultivation as a single-cropping rice variety in the rice-growing areas south of the Qiantang River in Zhejiang Province.

24. Lingliangyou 268

The two-line indica hybrid rice bred by Hunan Yahua Seed Science Research Institute was recognized as a super rice variety in 2011. It is suitable for double-cropping early-season rice planting in the middle and lower reaches of the Yangtze River, with an average growth period of 112.2 days. Resistance: comprehensive index of rice blast disease is 5.3, with the highest level of blast loss rate at level 7, and a resistance frequency of 90%; average level of bacterial leaf blight disease is 6, with the highest level at 7; level 3 of brown planthopper resistance and level 3 of white-backed planthopper resistance. The main indicators of rice quality are a head rice rate of 66.5%, a length-to-width ratio of 3.2, a chalky rice rate of 39%, a chalkiness degree of 4.4%, a gel consistency of 79 mm, and an amylose content of 12.3%. It participated in the regional trials of late-maturing early-season indica rice varieties in the middle and lower reaches of the Yangtze River in 2006 and 2007, with an average yield of 519.7 kg/*mu*. In the production trials in 2007, the average yield was 514.1 kg/*mu*. This variety is suitable for early-season rice planting in Jiangxi, Hunan, the northern part of Fujian, and the central and southern parts of Zhejiang rice-growing areas.

25. Huiliangyou No. 6

The two-line indica hybrid rice bred by the Rice Research Institute of Anhui Academy of Agricultural Sciences was recognized as a super rice variety in 2011. It is suitable for planting in the middle and lower reaches of the Yangtze River as a single-cropping rice, with an average growth period of 135.1 days. Resistance: comprehensive index of rice blast disease 5.7, highest level of panicle blast damage rate 9, level 7 of bacterial leaf blight, level 9 of brown planthopper, highly susceptible to rice blast disease and brown planthopper, susceptible to bacterial leaf blight, moderate heat tolerance during heading stage. Main rice quality indicators: a head rice rate of 58.8%, a length-width ratio of 2.9, a chalky grain rate of 33%, a chalkiness degree of 6.9%, a gel consistency of 76 mm, and an amylose content of 14.7%. It has participated in the regional trials for the mid-season late-maturing indica rice varieties in the middle and lower reaches of the Yangtze River for two years in 2009 and 2010, with an average yield of 578.0 kg/*mu*. In the production trial in 2011, the average yield was 604.2 kg/*mu*. This variety is suitable for planting in the rice areas of the Yangtze River Basin in Jiangxi, Hunan (except for the Wuling Mountain area), Hubei (except for the Wuling Mountain area), Anhui, Zhejiang, Jiangsu, and the northern part of Fujian and the southern part of Henan.

26. Teyou 582

The three-line indica hybrid rice bred by the Rice Research Institute of Guangxi Academy of Agricultural Sciences was recognized as a super rice variety in 2011. This variety is planted as early-season rice in southern Guangxi with a total growth period of 124 days. The brown rice rate is 82.3%, the head rice rate is 70.9%, the length-width ratio is 2.2, the chalky rice rate is 96%, the chalkiness degree is 23.8%, the gel consistency is 38 mm, and the amylose content is 21.6%; Resistance: level 5 for seedling leaf blast, level 9 for panicle blast, 42.8% loss rate for panicle blast, 6.5 comprehensive index for rice blast disease resistance, and level 7 for bacterial leaf blight pathogenicity type IV and level 9 for type V. It participated in the regional trials of early-season late-maturing rice in southern Guangxi for two years in 2007 and 2008, with an average yield of 531.01 kg/*mu*. It was also tested and demonstrated in Beiliu, Pingnan, Qinzhou and other places in 2007–2008, with an average yield of 564.7 kg/*mu*. This variety is suitable for planting as early-season rice in the rice-growing areas of southern and central regions in Guangxi, and attention should be paid to the prevention and control of diseases and pests such as rice blast disease and bacterial leaf blight.

27. Zhunliangyou 608

The two-line Indica hybrid rice bred by Hunan Longping High-Tech Seed Co. Ltd. was recognized as a super hybrid rice variety in 2012. It is suitable for double-cropping late-season rice cultivation in the middle and lower reaches of the Yangtze River, with an average growth period of 119.0 days. Its resistance levels include a comprehensive disease index of 5.2 for rice blast, the highest level of susceptibility to panicle blight of level 9, and level 9 resistance to bacterial leaf blight and brown planthopper. The key indicators of grain quality include a head rice rate of 51.6%, a length-to-width ratio of 3.2, a chalky rice rate of 11%, a chalkiness degree of 1.1%, a gel consistency of 55 mm, and an amylose content of 20.9%. In the regional trials conducted in the middle and lower reaches of the Yangtze River for the late-season mid-late-maturing indica rice group in 2007 and 2008, the average yield was 520.34 kg/*mu*. In the production trials conducted in 2008, the average yield reached 535.61 kg/*mu*. This variety is suitable for late-season rice cultivation in the central and northern parts of Guangxi, northern parts of Guangdong, central and northern parts of Fujian, central and southern parts of Jiangxi, central and southern parts of Hunan, and southern parts of Zhejiang.

28. Shenliangyou 5814

The two-line indica hybrid rice variety bred by the Tsinghua Shenzhen Longgang Research Institute of the National Hybrid Rice Engineering Technology Research Center was recognized as a super rice variety in 2012. It is suitable for single-cropping mid-season rice cultivation in the upstream rice-growing regions of the Yangtze River, with an average growth period of 158.7 days. Its resistance levels include comprehensive disease indices for rice blast of 6.5 and 5.0 in two years, respectively, with the highest level of susceptibility to panicle blight being level 7 and a level 9 resistance to brown planthopper. It is susceptible to rice blast and highly susceptible to brown planthoppers. The key indicators of grain quality include a head rice rate of 63.4%, a length-to-width ratio of 2.9, a chalky rice rate of 22%, a chalkiness degree of 2.9%, a gel consistency of 83 mm, and an amylose content of 17.8%, meeting the national standard of grade 2 high-quality rice according to *High Quality Paddy*. In the regional trials conducted for the mid-season late-maturing indica rice group in the upstream regions of the Yangtze River in 2014 and 2015, the average yield was 623.9 kg/*mu*. In the production trials conducted in 2016, the average yield reached 598.4 kg/*mu*. This variety is suitable for mid-season rice cultivation in the hilly rice-growing areas of Pingba, Sichuan Province, the low- and mid-altitude indica rice regions of Yunnan Province (excluding the Wuling Mountain area), and Guizhou Province (excluding the Wuling Mountain area), and areas with an altitude below 800 m in altitude in Chongqing Municipality (excluding the Wuling Mountain area) and the southern rice-growing areas of Shaanxi Province. It is not suitable for planting in areas with a high incidence of rice blast.

29. Guangliangyouxiang 66

The two-line indica hybrid rice bred by the Hubei Agricultural Technology Extension Center was recognized as a super rice variety in 2012. It is suitable for single-cropping mid-season rice cultivation in the middle and lower reaches of the Yangtze River, with an average growth period of 138.8 days. Resistance levels include a comprehensive disease index for rice blast of 5.1, with the highest level of susceptibility to panicle blight being level 7, level 5 resistance to bacterial leaf blight, level 7 resistance to brown planthopper, and medium susceptibility to both rice blast and bacterial leaf blight. The key indicators of grain quality include a head rice rate of 58.6%, a length-to-width ratio of 2.9, a chalky rice rate of 17.7%, a chalkiness degree of 3.4%, a gel consistency of 81 mm, and an amylose content of 16.9%, meeting the national standard of grade 3 high-quality rice according to *High Quality Paddy*. In the regional trials conducted for the late-maturing mid-season indica rice group in the middle and lower reaches of the Yangtze River in 2009 and 2010, the average yield was 555.1 kg/*mu*. In the production trials conducted in 2011, the average yield reached 553.5 kg/*mu*. This variety is suitable for mid-season rice cultivation in the rice-growing areas of Jiangxi, Hunan (excluding the Wuling Mountain area), Hubei (excluding the Wuling Mountain area), the central-southern part of Anhui, the Yangtze River Basin in Zhejiang, and the northern part of Fujian and the southern part of Henan.

30. Tianyou Huazhan

The three-line indica hybrid rice bred by the Hubei Agricultural Technology Extension Center was recognized as a super rice variety in 2012. It is suitable for double-cropping early-season rice cultivation in South China, with an average growth period of 123.1 days. The resistance levels include a comprehensive disease index for rice blast of 3.6, with the highest level of susceptibility to panicle blast being level 5, a level 7 resistance to bacterial leaf blight, a level 7 resistance to brown planthopper, a level 3 resistance to white-backed planthopper, a medium susceptibility to rice blast, a susceptibility to bacterial leaf blight and brown planthopper, and a medium resistance to white-backed planthopper. The key indicators of grain quality include a head rice rate of 63.0%, a length-to-width ratio of 2.8, a chalky rice rate of 20%, a chalkiness degree of 4.5%, a gel consistency of 70 mm, and an amylose content of 20.8%, meeting the national standard of grade 3 high-quality rice according to *High Quality Paddy*. In the regional trials conducted for the early-season indica rice group in South China in 2009 and 2010, the average yield was 502.5 kg/*mu*. In the production trials conducted in 2011, the average yield reached 502.8 kg/*mu*. This variety is suitable for early-season rice cultivation in double-cropping areas with a low incidence of bacterial leaf blight in Guangdong, central and southwestern Guangxi, southern Hainan, and the central-southern part of Jiangxi. It is suitable for single-cropping mid-season rice cultivation in fields with medium fertility in the rice-growing areas of Jiangxi, Hunan (excluding the Wuling Mountain area), Hubei (excluding the Wuling Mountain area), Anhui, Zhejiang, Yangtze River Basin rice-growing area in Jiangsu, the northern part of Fujian, the southern part of Henan, the areas with a low incidence of bacterial leaf blight in Yunnan and Guizhou (excluding the Wuling Mountain area), the low- and mid-altitude indica rice areas of Chongqing (excluding the Wuling Mountain area), the Sichuan Pingba hilly rice area, and the southern part of Shaanxi. In the late-season rice-growing areas of central-northern Guangxi, northern Guangdong, central-northern Fujian, central-southern Jiangxi, central-southern Hunan, and southern Zhejiang, it is suitable for cultivation.

31. Yiyou 673

The three-line indica hybrid rice variety developed by the Fujian Academy of Agricultural Sciences was recognized as a super rice variety in 2012. It is suitable for single-cropping mid-season rice cultivation in fields with medium fertility in the Yangtze River Basin, including Jiangxi, Hunan, Hubei, Anhui, Zhejiang, and Jiangsu, and the northern part of Fujian, areas below 1,300 m above sea level in Yunnan, southern Henan, and other areas with a low incidence of bacterial leaf blight. The average growth period is 133.8 days. The resistance levels include a comprehensive disease index for rice blast diseases of 4.1, with the highest level of susceptibility to panicle blast being level 9, a level 9 resistance to bacterial leaf blight, and a level 7 resistance to brown planthopper. The key indicators of grain quality include a head rice rate of 49.8%, a length-to-width ratio of 3.1, a chalky rice rate of 52%, a chalkiness degree of 6.7%, a gel consistency of 66 mm, and an amylose content of 16.4%. Besides, in the regional trials conducted for the mid-season late-maturing indica rice group in the Yangtze River Basin in 2006 and 2007, the average yield was 567.54 kg/*mu*. In the production trials conducted in 2008, the average yield reached 571.57 kg/*mu*. This variety is suitable for single-cropping mid-season rice cultivation in the Yangtze River Basin, including Jiangxi, Hunan, Hubei, Anhui, Zhejiang, Jiangsu, and the northern part of Fujian, areas below 1,300 m above sea level in Yunnan, and the southern part of Henan.

32. Shenyou 9516

The three-line indica hybrid rice variety developed by the Graduate School at Tsinghua University, Shenzhen was recognized as a super rice variety in 2012. It has a late-maturing growth period of 112 to 116 days. The grain quality is classified as national standard grade 3 high-quality rice according to *High Quality Paddy* and provincial standard grade 3 high-quality rice. The head rice rate ranges from 70.2% to 70.8%, the chalky rice rate from 10% to 46%, the chalkiness degree from 1.8% to 20.0%, the amylose content from 15.3% to 15.4%, the gel consistency of 70 to 80 mm, the length-to-width ratio of 3.2, and the taste quality scores of 79 to 80. It exhibits resistance to rice blast, with an overall resistance frequency of 88.5%. The resistance frequencies for subgroups B and C are 84.4% and 91.7%, respectively. In disease nursery evaluation, it is classified as level 1.5 for leaf blast and level 2.0 for panicle blast, showing moderate susceptibility to bacterial leaf blight. In the late-maturing trials conducted in 2008 and 2009, the average yield per *mu* was 518.5 kg and 480.51 kg, respectively. In the production trial in 2009, the average yield per *mu* was 446.86 kg. This variety is suitable for early and late planting in rice-growing areas outside of northern Guangdong Province, with attention to be paid to the prevention and control of bacterial leaf blight.

33. Y Liangyou 087

The two-line indica hybrid rice variety bred by the Ward Crop Research Institute in Nanning was recognized as a super rice variety in 2013. It is suitable for early-season rice planting in southern Guangxi, with a total growth period of approximately 128 days. The main quality indicators include a brown rice rate of 78.9%, a head rice rate of 63.7%, a length-to-width ratio of 2.8, a chalky rice rate of 24%, a chalkiness degree of 2.5%, a gel consistency of 72 mm, and an amylose content of 14.5%. In terms of resistance, it has a rating of level 5 for seedling blast, levels 3 to 7 for panicle blast, with the highest level being 7, and panicle blast loss ranging from 7.5% to 38.9%. The comprehensive resistance index for rice blast ranges from 4.5 to 6.5. It shows susceptibility of type IV at levels 5 to 7 for bacterial leaf blight and type V at levels 7 to 9. In the regional trials for early-season late-maturing rice in southern Guangxi in 2008 and 2009, the average yield per *mu* was 538.78 kg. In the production trial in 2009, the average yield per *mu* was 536.98 kg. This variety is suitable for early or late planting according to the local conditions in rice-growing areas outside of northern Guangdong Province, and for early rice planting in the southern Guangxi rice-growing area. Farmers in the Guangxi rice-growing area should pay attention to the prevention and control of rice blast, bacterial leaf blight, and other diseases and pests.

34. Tianyou 3618

The three-line indica hybrid rice variety bred by the Rice Research Institute of Guangdong Academy of Agricultural Sciences was recognized as a super rice variety in 2013. It has a total growth period of 126–127 days for early planting. While its rice quality does not reach the high-quality standard, its head rice rate ranges from 53.4% to 61.2%, the chalky rice rate ranges from 38% to 45%, the chalkiness degree ranges from 20.0% to 23.6%, and the amylose content ranges from 21.4% to 22.6%. The gel consistency is 52–55 mm, the length-to-width ratio is 3.2–3.3, and the taste quality is rated at 74–75 points. It is resistant to rice blast disease, with a resistance frequency of 95.7% for the entire population, and 96.0% and 92.1% for subgroups B and C, respectively. Field monitoring results showed resistance to leaf blast and moderate resistance to panicle blast, but moderate susceptibility to bacterial leaf blight in the C4 and C5 groups. In the regional trials for early planting in 2007 and 2008, the average yield per *mu* was 462.1 kg and 471.9 kg, respectively. In the production trial for early planting in 2008, the average yield per *mu* was 474.4 kg. This variety is suitable for early and late planting in rice-growing areas outside northern Guangdong Province.

35. Zhong 9 You 8012

The three-line indica hybrid rice variety bred by the China National Rice Research Institute was recognized as a super rice variety in 2013. It has a total growth period of 133.1 days on average as a single-cropping mid-season rice planting in the middle and lower reaches of the Yangtze River. The resistance indices are 5.4 for rice blast disease, the highest loss level for panicle blast reaching level 9, level 7 for bacterial leaf blight, and level 9 for brown planthopper. Its main rice quality indicators are a head rice rate of 55.5%, a length-to-width ratio of 3.1, a chalky rice rate of 31%, a chalkiness degree of 6.3%, a gel consistency of 69 mm, and an amylose content of 25.6%. In the regional trials for mid-season late-maturing indica rice varieties in the middle and lower reaches of the Yangtze River in 2006 and 2007, the average yield per *mu* was 567.51 kg. In the production trial in 2008, the average yield per *mu* was 558.47 kg. This variety is suitable for single-cropping mid-season rice planting in the Yangtze River Basin (except for the Wuling Mountain area) in Jiangxi, Hunan, Hubei, Anhui, Zhejiang, Jiangsu provinces, and in the northern part of Fujian and the southern part of Henan Province.

36. H You 515

The three-line indica hybrid rice variety bred by Hunan Agricultural University and Hengyang Agricultural Science Research Institute was recognized as a super rice variety in 2013. It has a total growth period of 112.9 days on average for double-cropping late-season rice planting in the middle and lower reaches of the Yangtze River. The resistance includes an index of 6.0 for rice blast disease, the highest loss level for panicle blast reaching level 9, level 7 for bacterial leaf blight, level 9 for brown planthopper, and it has medium cold tolerance during the heading stage. It is susceptible to rice blast disease, bacterial leaf blight, and brown planthopper. The main rice quality indicators include a 57.2% head rice rate, a length-to-width ratio of 3.5, a chalky rice rate of 25%, a chalkiness degree of 5.0%, a gel consistency of 56 mm, and an amylose content of 21.6%, meeting the national standard of grade 3 high-quality rice according to *High Quality Paddy*. In the regional trials for late-season early-maturing indica rice varieties in the middle and lower reaches of the Yangtze River in 2009 and 2010, the average yield per *mu* was 499.6 kg. In the production trial in 2010, the average yield per *mu* was 486.4 kg. This variety is suitable for late-season rice planting in the southern rice-growing areas of Jiangxi, Hunan, Hubei, Zhejiang, and Anhui provinces south of the Yangtze River.

37. Yongyou 15

The three-line hybrid rice of indica-japonica hybridization bred by the Crop Research Institute of Ningbo Academy of Agricultural Sciences and Ningbo Seed Co. Ltd. was recognized as a super rice variety in 2013. The total growth period in the central rice-growing region of Fujian Province averaged 147.7 days over two years. The comprehensive evaluation of resistance to rice blast disease over two years indicated medium susceptibility. The quality testing results are as follows: a brown rice rate of 80.1%, a milled rice rate of 72.2%, a head rice rate of 63.3%, an average grain length of 6.4 mm, a length-to-width ratio of 2.4, a chalky rice rate of 14%, a chalkiness degree of 2.2%, a transparency level of 1, an alkali digestion value of 3.8, a gel consistency of 84 mm, an amylose content of 15.3%, and a protein content of 7.7%. In the regional trials for mid-season late-maturing indica rice in Fujian Province in 2010 and 2011, the average yield per *mu* increased by 1.51% compared to the control group. In the production trial for mid-season rice in Fujian Province in 2012, the average yield per *mu* was 622.93 kg. This variety is suitable for single-cropping rice planting in Zhejiang Province and also suitable for mid-season rice planting in Fujian Province.

38. Y Liangyou No. 2

The two-line indica hybrid rice variety bred by Hunan Hybrid Rice Research Center was recognized as a super rice variety in 2014. It has an average total growth period of 139.1 days when grown as a single-cropping mid-season rice in the middle and lower reaches of the Yangtze River. First, the resistance levels are as follows: the comprehensive index for rice blast disease is 5.5, with the highest loss rate at level 9, bacterial leaf blight disease at level 7, and brown planthopper at level 9. It is highly susceptible to rice blast disease, bacterial leaf blight disease, and brown planthopper. Second, the main quality indicators for the rice are as follows: a head rice rate of 64.7%, a length-to-width ratio of 3.0, a chalky rice rate of 28%, a chalkiness degree of 3.6%, a gel consistency of 84 mm, and an amylose content of 15.5%, meeting the national standard of grade 3 high-quality rice according to *High Quality Paddy*. In the regional trials for mid-season late-maturing indica rice in the middle and lower reaches of the Yangtze River in 2011 and 2012, the average yield per *mu* was 615.4 kg. In the production trial in 2012, the average yield per *mu* was 580.4 kg. This variety is suitable for planting as a single-cropping mid-season rice variety in the Yangtze River Basin in Jiangxi, Hunan (except for the Wuling Mountain area), Hubei (except for the Wuling Mountain area), Anhui, Zhejiang, and Jiangsu, and in the northern part of Fujian and the southern part of Henan. It is not suitable for being planted in areas with a high incidence of rice blast disease.

39. Y Liangyou 5867

The two-line indica hybrid rice variety bred by Jiangxi Keyuan Seed Industry Co. Ltd. and other institutions was recognized as a super rice variety in 2014. It has an average total growth period of 137.8 days when grown as a single-cropping mid-season rice in the middle and lower reaches of the Yangtze River. The resistance levels are as follows: comprehensive index for rice blast disease is 4.0 with the highest loss rate at level 5, bacterial leaf blight disease at level 3, and brown planthopper at level 9. It is moderately resistant to rice blast disease, moderately resistant to bacterial leaf blight disease, highly susceptible to brown planthopper, and has average heat tolerance during the heading stage. The main quality indicators for the rice are as follows: a head rice rate of 64.9%, a length-to-width ratio of 3.0, a chalky rice rate of 25.3%, a chalkiness degree of 4.4%, a gel consistency of 73 mm, and an amylose content of 15.3%, meeting the national standard of grade 3 high-quality rice according to *High Quality Paddy*. In the regional trials for mid-season late-maturing indica rice in the middle and lower reaches of the Yangtze River in 2009 and 2010, the average yield per *mu* was 577.7 kg. In the production trial in 2011, the average yield per *mu* was 600.8 kg. This variety is suitable for planting as a single-cropping mid-season rice in the Yangtze River Basin in Jiangxi, Hunan (except for the Wuling Mountain area), Hubei (except for the Wuling Mountain area), Anhui, Zhejiang, Jiangsu, and in the northern part of Fujian and the southern part of Henan.

40. Liangyou 038

The two-line indica hybrid rice variety bred by Jiangxi Tianya Seed Industry Co. Ltd. was recognized as a super rice variety in 2014. It has an average total growth period of 122.6 days. The percentage of rough rice is 80.6%, the milled rice is 71.8%, the head rice is 59.5%, the grain length is 7.1 mm, the length-to-width ratio is 3.2, the chalky rice rate is 60%, the chalkiness degree is 6.0%, the amylose content is 21.4%, and the gel consistency is 78 mm. According to the natural induction identification for resistance to rice blast disease, it shows high susceptibility to rice blast disease with a rating of level 9 for panicle blast. In the regional trials in Jiangxi Province in 2008 and 2009, the average yield per *mu* was 569.74 kg. This variety is suitable for cultivation in areas of Jiangxi Province with a low incidence of rice blast disease.

41. C Liangyou Huazhan

The two-line indica hybrid rice variety bred by Hunan Golden Agricultural Science and Technology Co. Ltd. was recognized as a super rice variety in 2014. It is suitable for double-cropping early-season rice cultivation in South China with an average total growth period of 123.3 days. Resistance: The comprehensive index for rice blast disease is 3.9 and 3.8, respectively, over the span of two years, with a maximum rating of level 7 for panicle blast loss, level 7 for bacterial leaf blight, level 9 for brown planthopper, and level 9 for white-backed planthopper. It is susceptible to rice blast disease, bacterial leaf blight, highly susceptible to brown planthopper, and highly susceptible to white-backed planthopper. The main quality indicators include a head rice yield of 50.8%, a length-to-width ratio of 3.1, a chalky rice rate of 18%, a chalkiness degree of 2.9%, a gel consistency of 85 mm, and an amylose content of 12.8%. In the regional trials for early-season indica rice in South China in 2013 and 2014, the average yield per *mu* was 508.4 kg. In the production trials in 2015, the average yield per *mu* was 537.0 kg. This variety is suitable for early-season rice cultivation in areas with a low incidence of rice blast disease, particularly in the central and southwestern Guangdong, southern Fujian, southern Guangxi, and Hainan rice-growing regions with a low incidence of rice blast disease.

42. Guangliangyou 272

The two-line indica hybrid rice variety bred by the Institute of Food Crops Research at Hubei Academy of Agricultural Sciences was recognized as a super rice variety in 2014. In the regional trials for mid-season rice varieties in Hubei Province from 2010 to 2011, the grain quality was determined by the Agriculture and Rural Affairs Ministry's Food Quality Supervision and Inspection Center (Wuhan). The results show a hulling rate of 78.9%, a head rice yield of 62.5%, a chalky rice rate of 13%, a chalkiness degree of 1.3%, an amylose content of 16.4%, a gel consistency of 77 mm, and a length-to-width ratio of 3.0. The major physicochemical indicators meet the national standard of grade 2 high-quality rice according to *High Quality Paddy*. The average yield per *mu* in the regional trials was 604.50 kg, and the average total growth period was 139.8 days. In terms of disease resistance evaluation, the comprehensive index for rice blast disease was 6.9, with a maximum rating of level 9 for panicle blast loss, level 5 for bacterial leaf blight, highly susceptible to rice blast disease, and moderately susceptible to bacterial leaf blight. This variety is suitable for mid-season rice cultivation in areas of Hubei Province outside the southwestern region with either no or a low incidence of rice blast disease.

43. Liangyou No. 6

The two-line indica hybrid rice variety bred by Hubei Jingchu Seed Industry Co. Ltd. was recognized as a hybrid rice variety in 2014. It is suitable for double-cropping early-season rice cultivation in the middle and lower reaches of the Yangtze River, with an average growth period of 112.7 days. Resistance: the comprehensive index of rice blast disease is 6.6, with the highest level of panicle blast loss at level 9, bacterial leaf blight disease at level 7, brown planthopper at level 9, and white-backed planthopper at level 9. It is highly susceptible to rice blast disease, brown planthopper, white-backed planthopper, and susceptible to bacterial leaf blight disease. The main grain quality indicators are a head rice rate of 58.2%, a length-to-width ratio of 3.2, a chalky rice rate of 19%, a chalkiness degree of 3.9%, a gel consistency of 58 mm, and an amylose content of 22.1%, meeting the national standard of grade 3 high-quality rice according to *High Quality Paddy*. In the regional trials for early-season late-maturing indica rice varieties in the middle and lower reaches of the Yangtze River in 2008 and 2009, the average yield per *mu* was 521.8 kg. In the production trials in 2010, the average yield per *mu* was 455.8 kg. This variety is suitable for early-season rice cultivation in Jiangxi, Hunan, northern Guangxi, northern Fujian, and central-southern Zhejiang rice growing areas.

44. Liangyou 616

The two-line indica hybrid rice variety bred by China National Seed Group Co. Ltd. Fujian Agricultural and Livestock Seed Industry Co. Ltd. was recognized as a super rice variety in 2014. The average growth period of the two-year regional trial was 143.0 days. The comprehensive evaluation of rice blast disease resistance was moderate, with the varieties in the trial plot of Hunagtan Town, Jiangle County, identified as susceptible to rice blast disease. The grain quality inspection results were a brown rice rate of 80.4%, a milled rice rate of 73.0%, a head rice rate of 64.9%, a grain length of 7.1 mm, a length-to-width ratio of 2.9, a chalky rice rate of 39.0%, a chalkiness degree of 8.5%, a transparency level of 1, an alkali spreading value of 5.3, a gel consistency of 86 mm, an amylose content of 15.6%, and a protein content of 7.6%. In the Fujian Province regional trials for mid-season rice in 2009, the average yield per *mu* was 630.87 kg; in the continuation of the trials in 2010, the average yield per *mu* was 604.86 kg. In 2011, it participated in the provincial mid-season rice production trials, with an average yield per *mu* of 620.2 kg. This variety is suitable for mid-season rice cultivation in the rice-growing areas of Fujian Province.

45. Wufengyou 615

The three-line indica hybrid rice bred by the Rice Research Institute of Guangdong Academy of Agricultural Sciences was recognized as a super rice variety in 2014. The average growth period for early planting was 129 days. The grain quality did not reach the national high-quality standard, with the head rice rates ranging from 51.1% to 68.7%, the chalky rice rates ranging from 58% to 63%, the chalkiness degrees ranging from 21.3% to 23.5%, the amylose content ranging from 12.1% to 13.0%, the gel consistency from 86 to 90 mm, the length-to-width ratio from 2.8 to 2.9, and the taste quality scores from 75 to 77. It exhibits moderate resistance to rice blast disease, with a resistance frequency of 92.86% to 100% for the entire population and frequencies of 81.25% to 100% and 100% for subgroups B and C, respectively. In disease nursery evaluation, it scored 1.4 to 2.5 (highest single point: 4.0) for leaf blast and 1.8 to 4.0 (highest single point: 7.0) for panicle blast. It is susceptible to bacterial leaf blight (grade 7 for type IV, grade 9 for type V). In the early planting trials in 2010 and 2011, the average yield per *mu* was 447.22 kg and 543.42 kg, respectively. In the provincial production trials for early planting rice in 2011, the average yield per *mu* was 478.19 kg. This variety is suitable for early and late planting in rice-growing areas outside the northern part of Guangdong Province.

46. Shengtaiyou 722

The three-line indica hybrid rice bred by Hunan Dongting High-Tech Seed Co. Ltd. was recognized as a super rice variety in 2014. It is mainly cultivated as late-season rice in Hunan Province, with an average growth period of 112.6 days. Its resistance capabilities include an average leaf blast level of 4.1, a panicle blast level of 7.3, and a comprehensive resistance index for rice blast disease of 5.3. It has moderate tolerance to low temperatures. In terms of grain quality, it has a brown rice rate of 81.9%, a milled rice rate of 70.8%, a head rice rate of 57.1%, a grain length of 7.3 mm, a length-to-width ratio of 3.6, a chalky rice rate of 28%, a chalkiness degree of 2.5%, a transparency level of 1, an alkali digestion value of 6.0, a gel consistency of 55 mm, and an amylose content of 18.2%. In the two years of participation in the provincial trials in Hunan in 2010 and 2011, the average yield per *mu* was 501.49 kg. This variety is suitable for double-cropping late-season rice cultivation in areas of Hunan Province where the rice blast disease incidence is low.

47. Nei 5 You 8015

The three-line indica hybrid rice bred by the China National Rice Research Institute and Zhejiang Agricultural Science and Technology Seed Co. Ltd. was recognized as a super rice variety in 2014. It is mainly cultivated as a single-cropping mid-season rice in the middle and lower reaches of the Yangtze River, with an average growth period of 133.1 days. Its resistance capabilities include a comprehensive index of 5.9 for rice blast disease, the highest panicle blast damage level of 9, a bacterial leaf blight level of 9, and a brown planthopper level of 9. The main indicators of grain quality include a head rice rate of 52.2%, a length-to-width ratio of 3.0, a chalky rice rate of 30%, a chalkiness degree of 4.4%, a gel consistency of 76 mm, and an amylose content of 15.8%, meeting the national standard of grade 3 high-quality rice according to *High Quality Paddy*. In the regional trials for the mid-season late-maturing indica varieties group in the middle and lower reaches of the Yangtze River in 2007 and 2008, the average yield per *mu* was 590.8 kg. In the production trial in 2009, the average yield per *mu* was 591.4 kg. This variety is suitable for single-cropping mid-season rice cultivation in Jiangxi, Hunan, Hubei, Anhui, Zhejiang, Jiangsu, and the northern part of Fujian and the southern part of Henan.

48. Rongyou 225

The three-line indica hybrid rice bred by the Rice Research Institute of Jiangxi Academy of Agricultural Sciences and the Rice Research Institute of Guangdong Academy of Agricultural Sciences was recognized as a super rice variety in 2014. It is mainly cultivated as a double-cropping late-season rice in the middle and lower reaches of the Yangtze River, with an average growth period of 116.5 days. Its resistance capabilities include a comprehensive index of 5.8 for rice blast disease, the highest panicle blast damage level of 9, a bacterial leaf blight level of 5, a brown planthopper level of 9, and a black-streaked dwarf disease incidence rate of 63%. It has high susceptibility to rice blast disease, black-streaked dwarf disease, and brown planthopper, with moderate susceptibility to bacterial leaf blight. The cold tolerance during the heading stage is weak. The main indicators of grain quality include a head rice rate of 61.2%, a length-to-width ratio of 3.0, a chalky rice rate of 14%, a chalkiness degree of 2.8%, a gel consistency of 57 mm, and an amylose content of 24.4%. In the regional trials for the early-maturing group of late-season indica rice in the middle and lower reaches of the Yangtze River in 2009 and 2010, the average yield per *mu* was 516.4 kg. In the production trial in 2011, the average yield per *mu* was 510.5 kg. This variety is suitable for late-season rice cultivation in the double-cropping rice areas of Jiangxi and Hunan but is not recommended for areas with severe rice blast disease.

49. F You 498

The three-line indica hybrid rice bred by the Rice Research Institute of Sichuan Agricultural University was recognized as a super rice variety in 2014. It is mainly cultivated as a single-cropping mid-season rice in the upper reaches of the Yangtze River, with an average growth period of 155.2 days. Its resistance capabilities include a comprehensive index of 6.0 for rice blast disease, the highest panicle blast damage level of 7, a brown planthopper level of 7, and weak heat tolerance. It is susceptible to rice blast disease and brown planthopper. The main indicators of grain quality include a head rice rate of 69.2%, a length-to-width ratio of 2.8, a chalky rice rate of 21%, a chalkiness degree of 4.9%, a gel consistency of 80.5 mm, and an amylose content of 23.5%. In the regional trials for the late-maturing group of mid-season indica rice in the upper reaches of the Yangtze River in 2008 and 2009, the average yield per *mu* was 621.3 kg. In the production trial in 2010, the average yield per *mu* was 582.6 kg. This variety is suitable for single-cropping mid-season rice cultivation in the low- and mid-altitude indica rice areas of Yunnan, Guizhou (except for the Wuling Mountain area), Chongqing (except for the Wuling Mountain area), and the light rice blast disease areas in the Pingba hilly rice area of Sichuan. It can also be cultivated as mid-season rice in mountainous areas below 600 m altitude with a low rice blast disease incidence in Hunan Province.

50. H Liangyou 991

The two-line indica hybrid rice bred by Guangxi Zhaohe Seed Co. Ltd. was recognized as a super hybrid rice variety in 2015. It has a total growth period of around 108 days for late-season rice cultivation in central and northern Guangxi. The main indicators of grain quality include a brown rice rate of 77.7%, a head rice rate of 65.3%, a grain length of 6.8 mm, a length-to-width ratio of 3.2, a chalky rice rate of 22%, a chalkiness degree of 4.2%, a gel consistency of 85 mm, and an amylose content of 13.9%. In terms of resistance, it has a level 5 rating for seedling leaf blast disease, panicle blast disease incidence ranging from 85.45% to 93.0%, and a loss rate from 15.9% to 32.9%, with a level rating from 5 to 7. The comprehensive resistance index for rice blast disease is between 6.0 and 7.0, indicating a moderate to susceptible level of resistance. For bacterial leaf blight, it has a rating of 5 to 7 for type IV and a rating of 9 for type V, indicating a moderate to highly susceptible level of resistance. In terms of yield, the average yield per *mu* in the regional trials for the late-season mid-maturing rice varieties group in the central and northern rice farming areas of Guangxi in 2009 and 2010 was 482.28 kg. In the production trial in 2010, the average yield per *mu* was 414.22 kg. This variety is suitable for early and late-season rice cultivation in the central rice farming area of Guangxi, late-season rice cultivation in the northern rice farming area, and early-season rice cultivation in the southern rice farming area of Guangxi, based on local conditions.

51. N Liangyou 2

The two-line indica hybrid rice bred by Changsha Nianfeng Seed Industry Co. Ltd. and Hunan Hybrid Rice Research Center was recognized as a super rice variety in 2015. It has an average total growth period of 141.8 days when cultivated as mid-season rice in Hunan Province. In terms of resistance, it has a rating of 5.50 for leaf blast disease, 7.00 for stem blast disease, and a comprehensive resistance index of 5.25 for rice blast disease. It also has a resistance rating of 5 for bacterial leaf blight and sheath blast disease. The tolerance to high and low temperatures is moderate. In terms of grain quality, it has a husking rate of 80.2%, a brown rice rate of 70.6%, and a head rice rate of 66.0%. The grain length is 6.6 mm, with a length-to-width ratio of 3.0. The chalky rice rate is 18%, the chalkiness degree is 1.4%, the transparency level is 2, the alkali digestion value is 4, the gel consistency is 90 mm, and the amylose content is 15.0%. In the regional trials in Hunan Province in 2011 and 2012, the average yield per *mu* was 635.85 kg. This variety is suitable for mid-season rice cultivation in hilly areas of Hunan Province.

52. Yixiangyou 2115

The three-line indica hybrid rice bred by the College of Agriculture of Sichuan Agricultural University was recognized as a super hybrid rice variety in 2015. It has an average total growth period of 156.7 days when cultivated as a single-cropping mid-season rice in the upper reaches of the Yangtze River. In terms of resistance, it has a comprehensive resistance index of 3.6 for rice blast disease, with a maximum rating of level 5 for panicle blast disease and a resistance frequency of 33.7%, and a rating of level 9 for brown planthopper. It is moderately susceptible to rice blast disease and highly susceptible to brown planthopper. The main indicators of grain quality are a head rice rate of 54.5%, a length-to-width ratio of 2.9, a chalky rice rate of 15.0%, a chalkiness degree of 2.2%, a gel consistency of 78 mm, and an amylose content of 17.1%, meeting the national standard of grade 2 high-quality rice according to *High Quality Paddy*. In the regional trials for mid-season late-maturing indica rice varieties in the upper reaches of the Yangtze River in 2010 and 2011, the average yield per *mu* was 603.9 kg. In the production trials in 2011, the average yield per *mu* was 623.3 kg. This variety is suitable for single-cropping mid-season rice cultivation in the low- and mid-altitude indica rice areas of Yunnan, Guizhou (excluding Wuling Mountainous areas), Chongqing (excluding Wuling Mountainous areas), the hilly rice areas of Pingba, Sichuan, and the southern rice areas of Shaanxi.

53. Shenyou 1029

The three-line indica hybrid rice bred by Jiangxi Modern Seed Industry Co. Ltd. was recognized as a super rice variety in 2015. It has an average total growth period of 118.4 days when cultivated as a double-cropping late-season rice in the middle and lower reaches of the Yangtze River. In terms of resistance, it has a comprehensive resistance index of 5.9 for rice blast disease, with a maximum rating of level 9 for panicle blast disease, and a level 9 for bacterial leaf blight and brown planthopper. It is highly susceptible to rice blast disease, bacterial leaf blight, and brown planthopper, with weak cold tolerance during the heading stage. The main indicators of grain quality are a head rice rate of 52.9%, a length-to-width ratio of 3.0, a chalky rice rate of 28%, a chalkiness degree of 4.5%, a gel consistency of 71 mm, and an amylose content of 16.5%, meeting the national standard of grade 3 high-quality rice according to *High Quality Paddy*. In the regional trials for the double-cropping late-season early-maturing indica rice group in the middle and lower reaches of the Yangtze River in 2010 and 2011, the average yield per *mu* was 502.2 kg. In the production trials in 2012, the average yield per *mu* was 502.6 kg. This variety is suitable for late-season rice cultivation in the double-cropping rice areas of Jiangxi, Hubei, Zhejiang, and Anhui. It is not suitable for cultivation in areas with heavy rice blast disease.

54. Yongyou 538

The three-line hybrid rice of indica-japonica hybridization bred by Ningbo Seed Limited Company was recognized as a super hybrid rice variety in 2015. It achieved an average yield of 718.4 kg/*mu* in the regional trials for single-cropping late-seasoning hybrid japonica rice in Zhejiang Province in 2011 and 2012, and an average yield of 755.0 kg/*mu* in the provincial production trials in 2012. Its total growth period averages 153.5 days. According to the resistance identification conducted by the Zhejiang Academy of Agricultural Sciences in 2011–2012, it has an average rating of 1.1 for leaf blast disease, 5.0 for panicle blast disease, with a loss rate of 8.3%. It also has a comprehensive resistance index of 3.7, 2.4 for bacterial leaf blight and 9.0 for brown planthopper. According to the testing conducted by the National Rice and Quality Supervision and Inspection Center in 2011–2012, it has an average head rice rate of 71.2%, a length-to-width ratio of 2.1, a chalky rice rate of 39%, a chalkiness degree of 7.7%, a transparency rating of level 2, a gel consistency of 70.5 mm, and an amylose content of 15.5%. All quality indicators meet the standards for edible rice of grade 4 issued by the Department of Edible Rice Varieties. This variety is suitable for single-cropping rice cultivation in the rice-growing areas of Zhejiang Province.

55. Chunyou 84

The three-line japonica hybrid rice bred by China National Rice Research Institute and Zhejiang Agriculture and Science Seed Industry Co. Ltd. was recognized as a super rice variety in 2015. It achieved an average yield of 685.9 kg/*mu* in the regional trials for single-cropping late-season japonica rice in Zhejiang Province in 2010 and 2011, and an average yield of 658.3 kg/*mu* in the provincial production trials in 2012. Its total growth period averages 156.7 days. According to the resistance identification conducted by the Institute of Plant Protection of Zhejiang Academy of Agricultural Sciences in 2010–2011, it has an average rating of 0.3 for leaf blast disease, 2.0 for panicle blast disease with a loss rate of 1.0% and a comprehensive resistance index of 1.7, 6.0 for bacterial leaf blight, and 9.0 for brown planthopper. According to the testing conducted by the National Rice and Quality Supervision and Inspection Center in 2010–2011, it has an average head rice rate of 65.9%, a length-to-width ratio of 2.0, a chalky rice rate of 71%, a chalkiness degree of 11.3%, a transparency rating of level 3, a gel consistency of 80 mm, and an amylose content of 16.8%. Quality indicators meet the standards for edible rice of grade 5 and grade 6 issued by the Department of Edible Rice Varieties, respectively. This variety is suitable for single-cropping late-season rice cultivation in the rice-growing areas of Zhejiang Province.

56. Zheyou 18

The three-line hybrid rice of indica-japonica hybridization bred by the Institute of Crop and Nuclear Technology Utilization of Zhejiang Academy of Agricultural Sciences was recognized as a super rice variety in 2015. It achieved an average yield of 662.1 kg/*mu* in the regional trials for single-cropping indica-japonica hybrid rice in Zhejiang Province in 2010 and 2011, and an average yield of 672.4 kg/*mu* in the provincial production trials in 2011. Its total growth period averages 153.6 days. According to the resistance identification conducted by the Zhejiang Academy of Agricultural Sciences in 2010–2011, it has an average rating of 2.4 for leaf blast disease, 3.5 for bacterial leaf blight, 8.0 for brown planthopper, and 4.5 for panicle blast disease with a loss rate of 7.8% and a comprehensive resistance index of 4.3. According to the testing conducted by the National Rice and Quality Supervision and Inspection Center in 2010–2011, it has an average head rice rate of 67.3%, a length-to-width ratio of 2.0, a chalky rice rate of 34.0%, a chalkiness degree of 5.0%, a transparency rating of level 3, a gel consistency of 70 mm, and an amylose content of 14.7%. Over the two years, all quality indicators meet the standards for edible rice of grade 6 and grade 4 issued by the Department of Edible Rice Varieties. This variety is suitable for single-cropping rice cultivation in the rice-growing areas of Zhejiang Province.

57. Huiliangyou 996

The two-line indica hybrid rice bred by Hefei Keyuan Agricultural Science Research Institute and the Rice Research Institute of Anhui Academy of Agricultural Sciences was recognized as a super rice variety in 2016. It is suitable for planting as a mid-season rice in the middle and lower reaches of the Yangtze River, with an average growth period of 132.4 days. Resistance: The comprehensive resistance index for rice blast disease is 5.3, the highest level of panicle blast disease loss rate is 9, the level of bacterial leaf blight is 7, and the level of brown planthopper is 9. It is highly susceptible to rice blast disease and brown planthopper, and susceptible to bacterial leaf blight. The main indicators of grain quality are a head rice rate of 61.0%, a length-to-width ratio of 2.8, a chalky rice rate of 40%, a chalkiness degree of 9.7%, a gel consistency of 79 mm, and an amylose content of 13.1%. In the regional trials for the late-maturing group of mid-season indica rice in the middle and lower reaches of the Yangtze River in 2009 and 2010, the average yield per *mu* was 575.8 kg. In the production trials in 2011, the average yield per *mu* was 602.6 kg. This variety is suitable for planting as a mid-season rice variety in the Yangtze River Basin in Jiangxi, Hunan (except for the Wuling Mountain region), Hubei (except for the Wuling Mountain region), Anhui, Zhejiang, Jiangsu, and the northern part of Fujian and the southern part of Henan; it is not suitable for planting in areas heavily affected by rice blast disease.

58. Shenliangyou 870

The two-line indica hybrid rice bred by Guangdong Zhaohua Seed Co. Ltd. and Shenzhen Zhaonong Agricultural Technology Co. Ltd. was recognized as a super rice variety in 2016. It has an average growth period of 117 days for late-season cultivation. The grain quality is evaluated as national standard grade 3 high-quality rice according to *High Quality Paddy* and provincial standard grade 3, with the head rice rate ranging from 61.4% to 68.5%, the chalky rice rate from 12% to 15%, the chalkiness degree from 1.2% to 1.6%, the amylose content from 13.7% to 15.4%, the gel consistency from 60 to 68 mm, the length-to-width ratio from 3.2 to 3.3, and the taste score of 76 to 80 points. It also shows high resistance against rice blast disease with a resistance frequency of 93.5% to 95.7% in the entire population, and resistance frequencies of 93.8% to 95.8% and 92.3% to 94.7% for subgroups B and C, respectively. In disease nursery evaluations, it is rated at 1.6 to 2.3 for leaf blast and 3.0 for panicle blast; it is susceptible to bacterial leaf blight (IV type strain at level 7, V type at level 7). In the regional trials for late-season cultivation in 2012 and 2013, the average yield per *mu* was 496.67 kg and 446.63 kg, respectively. In the production trials in 2013, the average yield per *mu* was 430.08 kg. This variety is suitable for early-season and late-season cultivation in rice-growing areas outside of northern Guangdong Province, with attention to be paid to the prevention and control of bacterial leaf blight.

59. Deyou 4727

The three-line indica hybrid rice bred by Rice and Sorghum Research Institute of Sichuan Academy of Agricultural Sciences and Crop Research Institute of Sichuan Academy of Agricultural Sciences was recognized as a super rice variety in 2016. It has an average growth period of 149.4 days. Its quality measurements include a brown rice rate of 80.9%, a head rice rate of 68.3%, a length-to-width ratio of 2.8, a chalky rice rate of 38%, a chalkiness degree of 4.2%, a gel consistency of 86 mm, an amylose content of 15.0%, and a protein content of 9.7%. Disease resistance identification for rice blast disease showed ratings of level 6, 4, 7, and 5 for leaf blast in 2011, and level 5, 5, 5, and 7 for leaf blast in 2012, while ratings for panicle blast were level 5, 7, 5, and 5 in 2011 and level 5, 5, 5, and 5 in 2012. In the regional trials conducted for the mid-season late-maturing indica rice areas in Sichuan Province in 2011 and 2012, the average yield per *mu* was 549.16 kg. In the production trials in 2013, the average yield per *mu* was 563.55 kg. This variety is suitable for cultivation in the hilly areas of Pingba County in Sichuan Province.

60. Fengtianyou 553

The three-line indica hybrid rice bred by the Rice Research Institute of Guangxi Academy of Agricultural Sciences was recognized as a super rice variety in 2016. It has an average growth period of 115 days for late planting. Quality measurements include a grade 3 rating based on national and provincial standards, with the head rice rate ranging from 61.7% to 63.3%, the chalky rice rate ranging from 6% to 9%, the chalkiness degree ranging from 0.8% to 1.4%, the amylose content ranging from 14.4% to 15.6%, the gel consistency ranging from 72 to 82 mm, the length-to-width ratio ranging from 3.4 to 3.5, and the taste score ranging from 84 to 87. It exhibits resistance to rice blast disease, with a frequency of resistance ranging from 81.82% to 88.24% for the entire population and 68.42% to 88.46% and 85.71% to 100% for subgroups B and C, respectively. Disease nursery identification showed ratings of 1.5 to 2.8 for leaf blast and 2.5 to 3.5 for panicle blast. It also exhibits high susceptibility to bacterial leaf blight disease (type IV rating of 7 to 9, and type V rating of 9). In regional trials conducted in 2014 for late planting, the average yield per *mu* was 481.08 kg. In the retest for late planting in 2015, the average yield per *mu* was 460.55 kg. In the production trials for late planting in 2015, the average yield per *mu* was 467.00 kg. This variety is suitable for late planting in rice-growing areas outside of northern Guangdong Province. Special attention should be given to the prevention and control of bacterial leaf blight during cultivation.

61. Wuyou 662

The three-line indica hybrid rice bred by Jiangxi Huinong Seed Company Limited and Rice Research Institute of Guangdong Academy of Agricultural Sciences was recognized as a super rice variety in 2016. It has an average growth period of 119.2 days. The brown rice rate is 80.5%, the milled rice rate is 73.9%, the head rice rate is 51.1%, the grain length is 7.1 mm, the length-to-width ratio is 3.0, the chalky rice rate is 72%, the chalkiness degree is 10.1%, the amylose content is 20.0%, and the gel consistency is 40 mm. Natural induction identification for resistance to rice blast disease shows a rating of 9 for panicle blast, indicating high susceptibility to the disease. In the regional trials for rice varieties conducted in Jiangxi Province in 2010 and 2011, the average yield per *mu* was 495.38 kg. This variety is suitable for planting in Jiangxi Province.

62. Jiyou 225

The three-line indica hybrid rice bred by the Rice Research Institute of Jiangxi Academy of Agricultural Sciences was recognized as a super rice variety in 2016. It has an average growth period of 116.8 days. The brown rice rate is 80.3%, the milled rice rate is 73.2%, the head rice rate is 63.4%, the grain length is 6.7 mm, the length-to-width ratio is 3.0, the chalky rice rate is 10%, the chalkiness degree is 0.6%, the amylose content is 20.7%, and the gel consistency is 50 mm. Natural induction identification for resistance to rice blast disease shows a rating of 9 for stem blast, indicating high susceptibility to the disease. In the regional trials for rice varieties conducted in Jiangxi Province in 2012 and 2013, the average yield per *mu* was 545.54 kg. Besides, this variety is suitable for planting in Jiangxi Province.

63. Wufengyou 286

The three-line indica hybrid rice bred by Jiangxi Modern Seed Co. Ltd. and China National Rice Research Institute was recognized as a super rice variety in 2016. It is suitable for double-cropping early rice cultivation in the middle and lower reaches of the Yangtze River, with an average growth period of 113.0 days. Resistance: The comprehensive index of rice blast disease is 5.6, with the highest panicle blast loss rate at level 9, the resistance against bacterial leaf blight at level 7, brown planthopper at level 9, white-backed planthopper at level 7; it is highly susceptible to rice blast disease, bacterial leaf blight, brown planthopper, and white-backed planthopper. The main indicators of grain quality include a head rice rate of 65.4%, a length-to-width ratio of 2.7, a chalky rice rate of 27%, a chalkiness degree of 3.2%, a gel consistency of 86 mm, and an amylose content of 13.9%. In the regional trials for early-season late-maturing indica rice varieties for two consecutive years in 2012 and 2013, the average yield per *mu* was 537.4 kg. In the production trial in 2014, the average yield per *mu* was 488.1 kg. This variety is suitable for early-season rice cultivation in the double-cropping rice areas of Jiangxi, Hunan, northern Guangxi, northern Fujian, and central and southern Zhejiang, but not recommended for planting in areas prone to rice blast disease.

64. Wuyouhang 1573

The three-line indica hybrid rice bred by the Super Rice Research and Development Center in Jiangxi Province was recognized as a super rice variety in 2016. The average growth period is 123.1 days. The percentage of brown rice rate is 80.8%, the milled rice rate is 71.0%, the head rice rate is 64.3%, the grain length is 6.2 mm, the length-to-width ratio is 2.8, the chalky rice rate is 17%, the chalkiness degree is 1.9%, the amylose content is 20.0%, and the gel consistency is 50 mm. Natural induction identification for resistance to rice blast disease shows a level 9 resistance for stem blast, and the variety is highly susceptible to rice blast disease. In the provincial trials for rice varieties for two consecutive years in 2012 and 2013, the average yield per *mu* was 561.05 kg. This variety is suitable for planting in areas with a low incidence of rice blast disease in Jiangxi Province.

65. Y Liangyou 900

The two-line indica hybrid rice bred by Biocentury Transgene (China) Co. Ltd. was recognized as a super rice variety in 2017. It is suitable for double-cropping late rice planting in the southern part of China, with an average growth period of 114.0 days. Resistance: Rice blast disease comprehensive index was 5.9 and 6.3 for two consecutive years, respectively, with the highest level of panicle blast loss at level 9, bacterial leaf blight disease at level 9, and brown planthopper at level 9. It is highly susceptible to rice blast disease, highly susceptible to bacterial leaf blight disease, highly susceptible to brown planthopper. Major rice quality indicators include a head rice rate of 66.3%, a length-to-width ratio of 3.0, a chalky rice rate of 11%, a chalkiness degree of 1.9%, a gel consistency of 65 mm, and an amylose content of 14.5%. In the regional trials of the southern photosensitive late-season indica group in 2013 and 2014, the average yield per *mu* was 511.9 kg. In the production trial of 2015, the average yield per *mu* was 527.6 kg. This variety is suitable for double-cropping late rice planting in Hainan, central and southwestern plains of Guangdong, southern Guangxi, and southern Fujian, with a low incidence of rice blast disease, and special attention should be paid to the prevention and control of rice blast disease, bacterial leaf blight, and rice planthopper.

66. Longliangyou Huazhan

The two-line indica hybrid rice bred by Yuan Longping High-Tech Agriculture Co. Ltd. was recognized as a super rice variety in 2017. It is suitable for double-cropping late-season rice planting in the southern part of China, with an average growth period of 115.0 days. Resistance: The rice blast disease comprehensive index was 3.7 for two consecutive years, with the highest level of panicle blast loss at level 3, bacterial leaf blight disease at level 7, and brown planthopper at level 9. It is moderately resistant to rice blast disease, susceptible to bacterial leaf blight disease, highly susceptible to brown planthopper. Major rice quality indicators include a head rice rate of 68.4%, a length-to-width ratio of 3.1, a chalky rice rate of 7%, a chalkiness degree of 0.9%, a gel consistency of 75 mm, and an amylose content of 13.6%. For single-cropping mid-season rice planting in the upper reaches of the Yangtze River, the average growth period is 157.9 days. Resistance: The rice blast disease comprehensive index was 2.8 for two consecutive years, with the highest level of spike blast loss at level 3, brown planthopper at level 7, moderately resistant to rice blast disease, and susceptible to brown planthopper. Major rice quality indicators are a head rice rate of 67.3%, a length-to-width ratio of 3.0, a chalky rice rate of 8%, a chalkiness degree of 1.3%, a gel consistency of 84 mm, and an amylose content of 16.6%, reaching the national standard of grade 2 high-quality rice according to *High Quality Paddy*. In the regional trials for the southern photosensitive late-season indica group in 2014 and 2015, the average yield per *mu* was 510.8 kg. In the production trial of 2016, the average yield per *mu* was 501.0 kg. In the regional trials of the mid-season late-maturing indica group in the upper reaches of the Yangtze River in 2014 and 2015, the average yield per *mu* was 625.7 kg. In the production trial of 2016, the average yield per *mu* was 594.6 kg. It is suitable for double-cropping late-season rice planting in Guangdong (excluding northern rice growing areas), southern Guangxi,

Hainan Province, and southern Fujian. This variety is suitable for single-cropping mid-season rice planting in the hilly rice areas of Pingba, Sichuan Province, Guizhou Province (excluding the Wuling Mountain area), the low- and mid-altitude indica rice areas of Yunnan Province, Chongqing Municipality, and the southern rice areas of Shanxi Province.

67. Shenliangyou 8386

The two-line indica hybrid rice bred by Guangxi Zhaohe Seeds Co. Ltd. was recognized as a super hybrid rice variety in 2017. It is suitable for early rice planting in southern Guangxi, with an average growth period of 128.8 days. Resistance: The variety has resistance against seedling and leaf blast disease at level 4–5, spike blast loss rate of 11.24%–31.19%, with the highest loss rate at level 7 and a comprehensive index of resistance to rice blast disease at 4.8–7.0, bacterial leaf blight at level 5. It is moderately resistant to rice blast disease, and susceptible to bacterial leaf blight. Major rice quality indicators include a brown rice rate of 78.1%, a head rice rate of 55.6%, a length-to-width ratio of 3.5, a chalky rice rate of 9%, a chalkiness degree of 0.8%, a gel consistency of 78 mm, and an amylose content of 11.5%. In the regional trials of early-season rice in 2013 and 2014, the average yield per *mu* was 569.9 kg. In the production trial of 2014, the average yield per *mu* was 535.8 kg. This variety is suitable for early rice planting in southern Guangxi.

68. Y Liangyou 1173

The two-line indica hybrid rice bred by the National Plant Space Breeding Engineering Technology Research Center (South China Agricultural University) and the Hunan Hybrid Rice Research Center was recognized as a super rice variety in 2017. It has an average growth period of 125 days for early planting. The rice quality did not reach the high-quality rice according to *High Quality Paddy* standards, with a head rice rate of 34.6%–43.5%, a chalky rice rate of 9%–16%, a chalkiness degree of 1.3%–2.6%, an amylose content of 12.1%–14.4%, a gel consistency of 70–88 mm, a length-to-width ratio of 3.2, and a taste quality score of 71–83. It is resistant to rice blast disease, with a resistance frequency of 94.3%–96.97% for the entire population, and frequencies of 94.44%–100% and 85.7%–100% for subgroup B and subgroup C, respectively. Disease nursery identification shows a leaf blast rating of 1.0–1.3 and a panicle blast rating of 3.0. It is susceptible to bacterial leaf blight disease (type IV at level 7, type V bacteria at levels 7–9). In the early planting trials in 2013 and 2014, the average yield per *mu* was 488.54 kg and 476.42 kg, respectively. In the provincial production trial in 2014, the average yield per *mu* was 468.62 kg. This variety is suitable for early and late planting in rice-growing areas outside the northern part of Guangdong Province.

69. Yixiang 4245

The three-line indica hybrid rice bred by Yibin Academy of Agricultural Sciences was recognized as a super rice variety in 2017. It has an average growth period of 159.2 days for one season of middle-season rice planting in the upper reaches of the Yangtze River. Its resistance includes a comprehensive index of 4.9 for rice blast disease, the highest level of panicle blast loss rate at grade 7, a resistance frequency of 63.6%, and a level 9 resistance to brown planthopper. It is susceptible to rice blast disease and highly susceptible to brown planthopper. The main indicators of rice quality are reflected in a head rice rate of 66.0%, a length-to-width ratio of 2.9, a chalky rice rate of 10.5%, a chalkiness degree of 1.7%, a gel consistency of 78 mm, and an amylose content of 17.0%, meeting the national standard of grade 2 high-quality rice according to *High Quality Paddy*. In the regional trials for the mid-season late-maturing indica rice group in the upper reaches of the Yangtze River in 2009 and 2010, the average yield per *mu* was 584.9 kg. In the production trial in 2011, the average yield per *mu* was 615.0 kg. This variety is suitable for single-cropping middle-season rice planting in areas of low- and mid-altitude indica rice region in Yunnan and Guizhou (excluding the Wuling Mountain area), hilly rice fields in Pingba of Sichuan Province, and light rice blast disease areas in southern Shanxi Province.

70. Jifengyou 1002

The three-line indica hybrid rice bred by the Rice Research Institute of Guangdong Academy of Agricultural Sciences and Golden Rice Seeds Co. Ltd. of Guangdong was recognized as a super rice variety in 2017. The growth period for late planting is 121 days on average, and the rice quality does not reach the high-quality rice according to the national standards of *High Quality Paddy*. The head rice rate ranges from 57.5% to 67.8%, the chalky rice rate ranges from 28% to 40%, the chalkiness degree ranges from 4.2% to 5.7%, the amylose content is 22.0%, the gel consistency ranges from 44 to 84 mm, and the length-to-width ratio ranges from 3.0 to 3.1, with the taste score of 75 to 77 points. It is highly resistant to rice blast disease, with a resistance frequency of 100% for the entire group, and 100% for subgroups B and C. It has a leaf blast rating of 2.0–2.5 and a panicle blast rating of 2.3–2.5. It is susceptible to bacterial leaf blight (type IV at levels 5–7, type V at level 7). In the late planting regional trials in Guangdong Province in 2011, the average yield per *mu* for late planting was 494.95 kg, and in the retest in 2012, the average yield per *mu* was 505.89 kg. In the production trial in 2012 for late planting, the average yield per *mu* was 552.06 kg. This variety is suitable for planting in the plain areas of the central-southern and southwestern rice-growing areas in Guangdong Province.

71. Wuyou 116

The three-line indica hybrid rice bred by Guangdong Modern Agricultural Group Co. Ltd. and Rice Research Institute of Guangdong Academy of Agricultural Sciences was recognized as a super rice variety in 2017. It has an average growth period of 114 days for late planting. The rice quality is classified as grade 3 high-quality rice according to the national and provincial standards, with a head rice rate ranging from 54.9% to 57.6%, a chalky rice rate ranging from 4% to 20%, a chalkiness degree ranging from 0.5% to 2.7%, an amylose content of 15.3%, a gel consistency ranging from 60 to 74 mm, a length-to-width ratio ranging from 2.8 to 2.9, and a taste rating of 81 to 85 points. It is resistant to rice blast disease, with a resistance frequency of 81.82% to 83.9% for the entire group, and 68.42% to 68.8% and 100% for subgroups B and C, respectively. The leaf blast rating is 2.0–3.0, and the panicle blast rating is 2.2–3.5. However, it is highly susceptible to bacterial leaf blight (type IV at levels 7–9, type V at levels 7–9). In the regional trials in 2013 and 2014 for late planting, the average yield per *mu* was 475.55 kg and 545.36 kg, respectively. In the production trial in 2014 for late planting, the average yield per *mu* was 511.99 kg. This variety is suitable for late planting in the northern rice-growing areas of Guangdong Province and early and late planting in the central-northern rice-growing areas. Besides, special attention should be given to the prevention and control of bacterial leaf blight during cultivation.

72. Yongyou 2640

The three-line hybrid rice of indica-japonica hybridization bred by Ningbo Seed Co. Ltd. was recognized as a super rice variety in 2017. The average growth period is 141 days. The highest blast loss rate is level 3, and the comprehensive resistance index for rice blast disease is 3.25 with a disease level of 3. It has moderate resistance to rice blast disease and susceptibility to bacterial leaf blight. It shows a resistance level of 3 to representative strains of Zhejiang 173, PX079, and JS49-6, and a resistance level of 5 to KS-6-6. Based on the comprehensive evaluation of fungal resistance identification in Putian City, it has a moderate susceptibility to rice blast disease. The results of the rice quality test are as follows: a brown rice rate of 82.2%, a milled rice rate of 73.8%, a head rice rate of 41.0%, a grain length of 5.4 mm, a length-to-width ratio of 2.2, a chalky rice rate of 10.0%, a chalkiness degree of 2.1%, a transparency level of 2, an alkali digestion value of 6.4, a gel consistency of 78 mm, an amylose content of 14.1%, and a protein content of 10.7%. It participated in the regional trials for japonica rice in southern Henan Province in 2015, with an average yield of 666.6 kg/*mu*. In the follow-up trial in 2016, the average yield was 664.2 kg/*mu*. It participated in the production trial for japonica rice in southern Henan Province in 2017, with an average yield of 594.8 kg/*mu*. This variety is suitable for being cultivated in the southern japonica rice-growing areas of Henan Province.

73. Longliangyou 1988

The two-line indica hybrid rice bred by Yuan Longping High-Tech Agriculture Co. Ltd. and Hunan Yahua Seed Industry Science Research Institute was recognized as a super rice variety in 2018. It is suitable for late-season rice cultivation in South China, with an average growth period of 118.0 days. Resistance: The comprehensive index of rice blast disease was 5.4 and 2.3 in two years, with a maximum loss rate of level 7 for stem blast, level 9 for bacterial leaf blight, and level 9 for brown planthopper. It is susceptible to rice blast disease, highly susceptible to bacterial leaf blight, and highly susceptible to brown planthopper. The main indicators of rice quality include a head rice rate of 62.1%, a length-to-width ratio of 3.0, a chalky rice rate of 23.0%, a chalkiness degree of 4.1%, a gel consistency of 70 mm, and an amylose content of 14.0%. For mid-season rice cultivation in the upper reaches of the Yangtze River, the average growth period is 154.6 days. Resistance: The comprehensive index of rice blast disease was 5.0 and 3.0 in two years, with a maximum loss rate of 7 for panicle blast and 9 for brown planthopper. It is susceptible to rice blast disease and highly susceptible to brown planthopper. The main indicators of rice quality include a head rice rate of 63.1%, a length-to-width ratio of 2.9, a chalky rice rate of 12.0%, a chalkiness degree of 2.8%, a gel consistency of 87 mm, and an amylose content of 15.8%, meeting the national standard of grade 3 high-quality rice according to *High Quality Paddy*. In the green channel regional trials for late-season indica rice in South China in 2015 and 2016, the average yield was 508.9 kg/*mu*. In the production trial in 2016, the average yield was 494.5 kg/*mu*. In the green channel regional trials for mid-season japonica rice in the upper reaches of the Yangtze River in 2015 and 2016, the average yield was 650.6 kg/*mu*. In the production trial in 2016, the average yield was 616.8 kg/*mu*. This variety is suitable for late-season rice cultivation in Guangdong (excluding northern Guangdong rice-growing areas), southern Fujian, southern Guangxi, and Hainan rice-growing areas. It is also suitable for low- and mid-altitude indica rice-growing areas in Yunnan and Guizhou (excluding the Wuling Mountain area), rice-growing areas with an altitude below 800 m in Chongqing (excluding the Wuling Mountain area), hilly rice-growing areas in Pingba of Sichuan Province, and southern Shanxi, where rice blast disease incidence is low. It is not suitable for cultivation in areas with a high incidence of rice blast disease.

74. Shenliangyou 136

The two-line indica hybrid rice bred by Hunan Danong Seed Industry Technology Co. Ltd. was recognized as a super rice variety in 2018. It is suitable for cultivating mid-season rice in the middle and lower reaches of the Yangtze River, with an average growth period of 138.5 days. The comprehensive index for rice blast disease was 5.9 and 5.6 in two years, with the highest stem blast loss rate at level 9. It also showed a resistance level of 5 against bacterial leaf blight and a level of 9 against brown planthopper. The main indicators for rice quality include a head rice rate of 62.0%, a length-to-width ratio of 3.1, a chalky rice rate of 23.0%, a chalkiness degree of 2.7%, a gel consistency of 64 mm, and an amylose content of 15.2%, meeting the national standard of grade 3 high-quality rice according to *High Quality Paddy*. In the regional trials for mid-season late-maturing indica rice in the middle and lower reaches of the Yangtze River in 2013 and 2014, the average yield was 638.2 kg/*mu*. In the production trial in 2015, the average yield was 683.8 kg/*mu*. This variety is suitable for single-cropping mid-season rice cultivation in the Yangtze River Basin in Jiangxi, Hunan (excluding the Wuling Mountain area), Hubei (excluding the Wuling Mountain area), Anhui, Zhejiang, Jiangsu, and the northern part of Fujian and the southern part of Henan, where rice blast disease incidence is low.

75. Jingliangyou Huazhan

The two-line indica hybrid rice bred by Yuan Longping High-Tech Agriculture Co. Ltd. was recognized as a super rice variety in 2018. It is suitable for cultivating mid-season rice in the middle and lower reaches of the Yangtze River, with an average growth period of 138.5 days. The comprehensive index for rice blast disease was 2.1 and 2.7 in two years, with the highest stem blast loss rate at level 3. It also shows resistance level 7 against bacterial leaf blight and brown planthopper. It exhibits moderate resistance to rice blast disease, susceptibility to bacterial leaf blight and brown planthopper. The main indicators for rice quality include a head rice rate of 66.4%, a length-to-width ratio of 3.1, a chalky rice rate of 13%, a chalkiness degree of 3.0%, a gel consistency of 81 mm, and an amylose content of 14.1%. In the Wuling Mountain area, where the full growth period for mid-season rice cultivation is 150.0 days, the comprehensive index for rice blast disease was 1.8 and 1.6 in two years, with the highest panicle blast loss rate at level 1. It exhibits resistance to rice blast disease and cold tolerance. The main indicators for rice quality include a head rice rate of 65.3%, length-to-width ratio of 3.1, chalky rice rate of 12%, chalkiness degree of 2.9%, gel consistency of 79 mm, and amylose content of 15.0%, meeting the national standard of grade 3 high-quality rice according to *High Quality Paddy*. In the regional trials for mid-season indica rice in the middle and lower reaches of the Yangtze River in 2014 and 2015, the average yield was 713.1 kg/*mu*. In the production trial in 2016, the average yield was 603.0 kg/*mu*. In the Wuling Mountain area, the average yield in the regional trials for middle-season indica rice in 2014 and 2015 was 619.16 kg/*mu*. In the production trial in 2016, the average yield was 578.40 kg/*mu*. This variety is suitable for single-cropping mid-season rice cultivation in the Yangtze River Basin in Jiangxi, Hunan (excluding the Wuling Mountain area), Hubei (excluding the Wuling Mountain area), Anhui, Zhejiang, Jiangsu, and the northern part of Fujian and the southern part of Henan, where rice blast disease incidence is low. It is also suitable for single-cropping mid-season rice cultivation in the Wuling Mountain area with an altitude below 800 m in Guizhou, Hunan, Hubei, and Chongqing.

76. Wuyou 369

The three-line indica hybrid rice bred by Hunan Taibang Agricultural Technology Co. Ltd. and Rice Research Institute of Guangdong Academy of Agricultural Sciences was recognized as a super rice variety in 2018. In the joint production test for the mid-maturing groups in the central and northern regions of Guangxi in 2017, the average yield per *mu* for early-season rice was 520.2 kg, and for late-season rice was 485.8 kg. The full growth period was 122.0 days for early-season rice and 107.6 days for late-season rice when planted in the central and north of Guangxi. This variety has moderate resistance to rice blast disease, with a comprehensive index of 5.5 and 4.5 in two consecutive years, and the highest panicle blast loss rate of 7 level, showing intermediate susceptibility to rice blast disease. For bacterial leaf blight, this variety demonstrated a resistance level of 5 to 7 (in early and late season in 2017), with a susceptibility range from intermediate to high. When planting this variety, you should pay more attention to the prevention and control of diseases such as rice blast. This variety is suitable for planting early- and late-season rice in the central and northern regions of Guangxi, and suitable planting seasons can be selected according to trial demonstrations of the variety and growth periods in other rice planting areas.

77. Neixiang 6 You 9

The three-line indica hybrid rice bred by the Rice and Sorghum Research Institute of Sichuan Academy of Agricultural Sciences was recognized as a super hybrid rice variety in 2018. When planted as a single-cropping mid-season rice crop in the upper reaches of the Yangtze River, the full growth period averaged 155.9 days. This variety has moderate resistance to rice blast disease, with a comprehensive index of 4.5, and the highest panicle blast loss rate of 7 levels. It has a high susceptibility to brown planthopper, with a rating of 9. The heat tolerance during the heading stage is moderate, and it is highly susceptible to rice blast disease and brown planthopper. The main indicators for the quality of rice include a head rice rate of 47.2%, a length-to-width ratio of 2.7, a chalky rice rate of 52%, a chalkiness degree of 10.2%, a gel consistency of 77 mm, and an amylose content of 21.1%. In the regional trials for the mid-season late-maturing indica rice group in the upper reaches of the Yangtze River in 2012 and 2013, the average yield per *mu* was 618.4 kg. In the production tests in 2014, the average yield per *mu* was 612.2 kg. This variety is suitable for planting single-cropping mid-season rice in the low- and mid-altitude indica rice areas of Yunnan, Guizhou (except Wuling Mountain area), Chongqing (except Wuling Mountain area), the hilly rice areas of Pingba in Sichuan, and the southern rice areas of Shanxi.

78. Shuyou 217

The three-line indica hybrid rice bred by the Rice Research Institute of Sichuan Agricultural University was recognized as a super hybrid rice variety in 2018. When planted as a single-cropping mid-season rice crop in the upper reaches of the Yangtze River, the full growth period averaged 155.4 days. This variety has moderate resistance to heat during the heading stage and high susceptibility to rice blast disease and brown planthopper. The comprehensive index of rice blast disease is 5.0, with the highest panicle blast loss rate of 7 levels and a rating of 9 for brown planthopper. The main indicators for the quality of rice include a head rice rate of 62.9%, a length-to-width ratio of 3.1, a chalky rice rate of 46%, a chalkiness degree of 7.1%, a gel consistency of 63 mm, and an amylose content of

21.4%. In the regional trials of the mid-season late-maturing indica rice groups in the upper reaches of the Yangtze River in 2012 and 2013, the average yield per *mu* was 624.2 kg. In the production tests in 2014, the average yield per *mu* was 618.1 kg. This variety is suitable for planting single-cropping mid-season rice in the low- and mid-altitude indica rice areas of Yunnan, Guizhou (except for the Wuling Mountain area), Chongqing (except for the Wuling Mountain area), the hilly rice areas of Pingba in Sichuan, and the southern rice areas of Shanxi. However, it should not be planted in areas with a high incidence of rice blast disease.

79. Luyou 727

The three-line indica hybrid rice bred by the Rice and Sorghum Research Institute of Sichuan Academy of Agricultural Sciences and the Crop Research Institute of Sichuan Academy of Agricultural Sciences was recognized as a super hybrid rice variety in 2018. When planted as a single-cropping mid-season rice crop in the upper reaches of the Yangtze River, the full growth period averaged 157.6 days. The comprehensive index for rice blast disease was 3.9 and 3.5 in two consecutive years, with the highest panicle blast loss rate at level 7, and a rating of 9 for brown planthopper. It shows weak heat tolerance during the heading stage and high susceptibility to rice blast disease and brown planthopper. The main indicators for rice quality include a head rice rate of 54.6%, a length-to-width ratio of 3.0, a chalky rice rate of 35%, a chalkiness degree of 5.6%, a gel consistency of 73 mm, and an amylose content of 21.1%. In the regional trials of the mid-season late-maturing indica rice groups in the upper reaches of the Yangtze River in 2013 and 2014, the average yield per *mu* was 627.5 kg. In the production tests in 2015, the average yield per *mu* was 623.0 kg. This variety is suitable for planting single-cropping mid-season rice in the low- and mid-altitude indica rice areas of Yunnan and Guizhou (except for the Wuling Mountain area), rice areas with an altitude below 800 m in Chongqing (except for the Wuling Mountain area), the hilly rice areas of Pingba in Sichuan, and the southern rice areas of Shanxi.

80. Jiyou 615

The three-line indica hybrid rice bred by the Rice Research Institute of Guangdong Academy of Agricultural Sciences and Golden Rice Seeds Co. Ltd. of Guangdong was recognized as a super rice variety in 2018. It has an average full growth period of 110 days for late planting. The rice quality is identified as grade 3 according to both national and provincial standards, with a head rice rate of 49.7% to 58.0%, a chalky rice rate of 11% to 39%, a chalkiness degree of 2.6% to 5.6%, an amylose content of 20.3% to 22.1%, a gel consistency of 50 to 54 mm, a length-to-width ratio of 3.2 to 3.3, and a taste quality of 70 to 78 points. It exhibits high resistance to rice blast disease, with a resistance frequency of 96.8% to 100% for the entire population and frequencies of 93.8% to 100% and 100% for subgroups B and C, respectively. In disease nursery identification, it is rated at levels 1.8 to 2.3 for leaf blast and 2.5 to 3.0 for panicle blast. It shows high susceptibility to bacterial leaf blight (ranging from level 7 to 9 for type IV bacteria and 7 to 9 for type V). In the late planting trials in 2013 and 2014, the average yield per *mu* was 470.48 kg and 525.50 kg, respectively. In the late planting production test in 2014, the average yield per *mu* was 522.36 kg. It is suitable for early and late planting in the plain areas of the central and northern rice-growing regions of Guangdong Province.

81. Wuyou 1179

The three-line indica hybrid rice bred by the National Plant Space Breeding Engineering Technology Research Center (South China Agricultural University) was recognized as a super rice variety in 2018. It has an average full growth period of 123.5 days for early planting. The rice quality does not reach the high-quality rice according to the standard of *High Quality Paddy*, with a head rice rate of 28.3% to 32.4%, a chalky rice rate of 11% to 13%, a chalkiness degree of 2.3% to 3.4%, an amylose content of 13.6% to 15.6%, a gel consistency of 80 to 90 mm, a length-to-width ratio of 2.7 to 2.8, and a taste score of 76 to 78. It exhibits high resistance to rice blast disease, with a resistance frequency of 91.18% to 100% for the entire population and frequencies of 84.21% to 100% and 100% for subgroups B and C, respectively. In disease nursery identification, it is rated at levels 2.0 to 2.5 for leaf blast and 3.0 to 3.5 for panicle blast. It shows high susceptibility to bacterial leaf blight (ranging from level 5 to 9 for type IV bacteria and 7 to 9 for type V). In the early planting trials in 2013, the average yield per *mu* was 477.19 kg, and in the retest in 2014, the average yield per *mu* was 512.03 kg. In the early planting production test in 2014, the average yield per *mu* was 494.35 kg. It is suitable for both early and late planting in the rice-growing regions of Guangdong Province.

82. Yongyou 1540

The three-line hybrid rice of indica-japonica hybridization bred by the Crop Research Institute of Ningbo Academy of Agricultural Sciences and Ningbo Seed Co. Ltd. was recognized as a super hybrid rice variety in 2018. It is suitable for single-cropping late-season rice planting in the middle and lower reaches of the Yangtze River, with an average full growth period of 151.0 days. The comprehensive index for rice blast disease is 5.6, with a maximum panicle blast loss rate of level 9, bacterial leaf blight rating at level 5, and brown planthopper rating at level 9. It is highly susceptible to rice blast disease, moderately susceptible to bacterial leaf blight, and highly susceptible to brown planthopper. It has a head rice rate of 70.2%, a length-to-width ratio of 2.3, a chalky grain rate of 18%, a chalkiness degree of 3.0%, a gel consistency of 75 mm, and a straight chain starch content of 14.3%. In the regional trials for single-cropping late-season japonica rice conducted in the middle and lower reaches of the Yangtze River in 2012 and 2013, the average yield per *mu* was 714.8 kg. In the production test in 2014, the average yield per *mu* was 683.7 kg. This variety is suitable for single-cropping late-season rice planting in Zhejiang, Shanghai, southern Jiangsu, and Hubei japonica rice regions but is not recommended for areas prone to rice blast disease.

83. Shenliangyou 862

The two-line indica hybrid rice bred by Jiangsu Mingtian Seed Technology Co. Ltd. was recognized as a super rice variety in 2019. It is suitable for single-cropping mid-season rice planting in the middle and lower reaches of the Yangtze River, with an average full growth period of 134.0 days. The comprehensive index for rice blast disease is 4.7, with a maximum panicle blast loss rate of level 7, bacterial leaf blight rating at level 5, and brown planthopper rating at level 7. It exhibits strong heat tolerance during the heading stage but is susceptible to rice blast disease, moderately susceptible to bacterial leaf blight, and susceptible to brown planthopper. The main rice quality indicators include a head rice rate of 53.5%, a length-to-width ratio of 3.2, a chalky rice rate of 14%, a chalkiness degree of 1.9%, a gel consistency of 76 mm, and an amylose content of 11.5%. In the regional trials for mid-season late-maturing japonica rice conducted in the middle and lower reaches of the Yangtze River in 2012 and 2013, the average yield per *mu* was 630.0 kg. In the production test in 2014, the average yield per *mu* was 607.9 kg. This variety is suitable for single-cropping mid-season rice planting in the rice areas of the Yangtze River Basin, including Jiangxi, Hunan (excluding Wuling Mountain area), Hubei (excluding Wuling Mountain area), Anhui, Zhejiang, Jiangsu, and the northern part of Fujian and the southern part of Henan. It is not recommended for areas prone to rice blast disease.

84. Longliangyou 1308

The two-line indica hybrid rice bred by Yuan Longping High-Tech Agriculture Co. Ltd. was recognized as a super rice variety in 2019. It is suitable for single-cropping mid-season rice planting in the middle and lower reaches of the Yangtze River, with an average full growth period of 137.2 days. The comprehensive index for rice blast disease in two years was 3.0 and 1.7 respectively, with a maximum panicle blast loss rate of level 3, bacterial leaf blight rating at level 5, and brown planthopper rating at level 9. It exhibits medium resistance to rice blast disease, moderate susceptibility to bacterial leaf blight, and high susceptibility to brown planthopper. The main rice quality indicators include a head rice rate of 61.4%, a length-to-width ratio of 3.1, a chalky rice rate of 10%, a chalkiness degree of 2.6%, a gel consistency of 63 mm, and an amylose content of 13.5%. In the regional trials conducted for mid-season late-maturing japonica rice in the green channel area in the middle and lower reaches of the Yangtze River in 2015 and 2016, the average yield per *mu* was 653.6 kg. In the production test in 2016, the average yield per *mu* was 597.2 kg. This variety is suitable for single-cropping rice planting in the Yangtze River Basin, including Jiangxi, Hunan (excluding Wuling Mountain area), Hubei (excluding Wuling Mountain area), Anhui, Zhejiang, Jiangsu, the northern part of Fujian, and the southern part of Henan. Attention should be paid to the prevention and control of rice blast disease in areas prone to its occurrence.

85. Longliangyou 1377

The two-line indica hybrid rice bred by Yuan Longping High-Tech Agriculture Co. Ltd. was recognized as a super rice variety in 2019. It has a full growth period of 155.1 days when planted as a mid-season rice crop in the upper reaches of the Yangtze River, with two-year comprehensive indexes for rice blast disease of 3.8 and 1.6, respectively. The highest level of panicle blast loss rate is at grade 5 and brown planthopper is at grade 9, with moderate susceptibility to rice blast disease and high susceptibility to brown planthopper. Its main rice quality indicators are a head rice rate of 63.4%, a length-width ratio of 3.0, a chalky rice rate of 13.0%, a chalkiness degree of 2.4%, a gel consistency of 76 mm, and an amylose content of 16.5%, meeting the national standard of grade 2 high-quality rice according to *High Quality Paddy*. When planted as a single-cropping mid-season rice crop in the middle and lower reaches of the Yangtze River, it has an average full growth period of 142.1 days, and two-year comprehensive indexes for rice blast disease of 2.8 and 2.7, respectively. It has the highest level of panicle blast loss rate at grade 5, bacterial leaf blight at grade 5, brown planthopper at grade 9, with moderate susceptibility to rice blast disease, moderate susceptibility to bacterial leaf blight, and high susceptibility to brown planthopper. Its main rice quality indicators are a head rice rate of 63.2%, a length-width ratio of 3.1, a chalky rice rate of 5%, a chalkiness degree of 0.6%, a gel consistency of 84 mm, and an amylose content of 15.1%, meeting the national standard of grade 3 high-quality rice according to *High Quality Paddy*. When planted as a late-season rice crop in southern China, it has an average full growth period of 119.5 days, and two-year comprehensive indexes for rice blast disease of 3.4 and 2.3, respectively, It has the highest level of rice blast loss rate at grade 3, bacterial leaf blight at grade 7, and brown planthopper at grade 9, with moderate resistance to rice blast disease, susceptibility to bacterial leaf blight, and high susceptibility to brown planthopper. The main rice quality indicators are a head rice rate of 66.2%, a length-width ratio of 3.1, a chalky rice rate of 22.3%, a chalkiness degree of 5.0%, a gel consistency of 70 mm, and an amylose content of 14.1%. In the two-year regional trials for mid-season late-maturing indica rice varieties in the upper reaches of the Yangtze River, the average yield was 655.2 kg/*mu* in the green channel trial area, and reached 626.9 kg/*mu* in the production trial in 2016. In the two-year regional trials for mid-season late-maturing indica rice varieties in the middle and lower reaches of the Yangtze River, the average yield was 657.1 kg/*mu* in the green channel trial area, and reached 589.9 kg/*mu* in the production trial in 2016. In the two-year regional trials of the green channel area for late-season indica rice varieties in southern China, the average yield was 491.3 kg/*mu*, and reached 472.1 kg/*mu* in the production trial in 2016. This variety is suitable for planting as a late-season rice crop in Guangdong (excluding the northern rice-growing area of Guangdong), southern Fujian, southern Guangxi, and Hainan; and for planting as a single-cropping mid-season rice crop, it can be grown in the Yangtze River Basin rice-growing areas in Hubei (excluding the Wuling Mountain area), Hunan (excluding the Wuling Mountain area), Jiangxi, Anhui, Jiangsu, the mid-season rice-growing areas in Zhejiang Province, the northern rice-growing area of Fujian Province, and the southern rice-growing area of Henan Province; and for planting as a mid-season rice crop, it can be grown in the low- and mid-altitude indica rice-growing areas in Yunnan and Guizhou (excluding the Wuling Mountain area), rice-growing areas with an altitude below 800 m in Chongqing (excluding the Wuling Mountain area), the hilly rice-growing area in Pingba, Sichuan, and the low-incidence area of rice blast disease in southern Shanxi. Be sure to pay attention to the prevention and control of rice blast disease in areas where it frequently occurs.

86. Heliangyou 713

The two-line indica hybrid rice bred by Guangxi Hengmao Agricultural Technology Co. Ltd. was recognized as a super rice variety in 2019. It is suitable for planting as a double-cropping late-season rice crop in South China, with an average full growth period of 114.3 days. The two-year comprehensive indexes for rice blast disease are 3.7 and 2.0, respectively, with the highest level of rice blast loss rate at grade 3, bacterial leaf blight at grade 5, and brown planthopper at grade 9. It has moderate resistance to rice blast disease, moderate susceptibility to bacterial leaf blight, and high susceptibility to brown planthopper. The main rice quality indicators include a head rice rate of 68.3%, a length-width ratio of 3.0, a chalky rice rate of 12%, a chalkiness degree of 1.0%, a gel consistency of 72 mm, and an amylose content of 14.5%. In the regional trials for the photosensitive late-season indica rice group in South China in 2015 and 2016, the average yield was 497.1 kg/*mu*; and in the production trial in 2016, the average yield was 502.3 kg/*mu*. This variety is suitable for planting as a late-season rice crop in the double-cropping regions in Guangdong (excluding the northern rice-growing area), southern Guangxi, Hainan, and the southern part of Fujian.

87. Y Liangyou 957

The two-line indica hybrid rice bred by Biocentury Transgene (China) Co. Ltd. and Hunan Yuanchuang Super Rice Technology Co. Ltd. was recognized as a super rice variety in 2019. It has an average growth period of 143.4 days when planted as single-cropping mid-season rice in the middle and lower reaches of the Yangtze River. The comprehensive index for blast disease over two years was 5.1 and 4.1, respectively, with the highest rating for panicle blast at level 9, bacterial leaf blight at level 5, and brown planthopper at level 9. It is highly susceptible to rice blast disease, moderately susceptible to bacterial leaf blight, and highly susceptible to brown planthopper. The main rice quality indicators include a head rice rate of 60.7%, a length-to-width ratio of 3.0, a chalky rice rate of 8%, a chalkiness degree of 1.2%, a gel consistency of 80 mm, and an amylose content of 16.3%, meeting the national standard of grade 2 high-quality rice according to *High Quality Paddy*. In regional trials conducted in the middle and lower reaches of the Yangtze River for the mid-season indica rice group from 2014 to 2015, the average yield per *mu* was 641.2 kg, while in production trials in 2016, the average yield per *mu* was 591.4 kg. This variety is suitable for single-cropping mid-season rice cultivation in the Yangtze River Basin in Hubei and Hunan provinces (excluding the Wuling Mountain area), Jiangxi, Anhui, and Jiangsu provinces. It can also be planted in the central rice-growing areas of Zhejiang Province, the northern portion of Fujian Province, and the southern part of Henan Province.

88. Longliangyou 1212

The two-line indica hybrid rice bred by Yuan Longping High-Tech Agriculture Co. Ltd. was recognized as a super rice variety in 2019. It has an average growth period of 140.4 days when planted as single-cropping mid-season rice in the middle and lower reaches of the Yangtze River. The comprehensive index for rice blast disease over two years was 2.9 and 3.1, respectively, with the highest rating for panicle blast at level 5, bacterial leaf blight at level 5, and brown planthopper at level 9. It is moderately susceptible to rice blast disease, moderately susceptible to bacterial leaf blight, and highly susceptible to brown planthopper. The main rice quality indicators are a head rice rate of 62.2%, a length-to-width ratio of 3.1, a chalky rice rate of 8%, a chalkiness degree of 0.9%, a gel consistency of 85 mm, and an amylose content of 15.0%, meeting the national standard of grade 3 high-quality rice according to *High Quality Paddy*. In the Wuling Mountain area, the average growth period for this variety is 149.1 days when planted as mid-season rice. The comprehensive index for rice blast disease over two years was 1.8 and 1.8, respectively, with the highest rating for panicle blast at level 1, indicating resistance to blast disease. It is moderately cold-tolerant. The main rice quality indicators include a head rice rate of 66.5%, a length-to-width ratio of 3.1, a chalky rice rate of 12%, a chalkiness degree of 3.8%, a gel consistency of 58 mm, and an amylose content of 15.5%, meeting the national standard of grade 3 high-quality rice according to *High Quality Paddy*. In regional trials conducted in the middle and lower reaches of the Yangtze River for the mid-season late-maturing indica rice group from 2015 to 2016, the average yield per *mu* was 660.8 kg; while in production trials in 2016, the average yield per *mu* was 654.1 kg. In the Wuling Mountain area, the average yield per *mu* in the regional trials for the mid-season indica rice group from 2015 to 2016 was 655.41 kg; while in production trials in 2016, the average yield per *mu* was 585.89 kg. This variety is suitable for single-cropping mid-season rice cultivation in the Yangtze River Basin (excluding the Wuling Mountain area), Jiangxi, Anhui, and Jiangsu provinces. It can also be cultivated in the central rice-growing areas of Zhejiang Province, the northern portion of Fujian Province, and the southern part of Henan Province. It is suitable for single-cropping mid-season rice cultivation in the Wuling Mountain area with an altitude below 800 m in Guizhou, Hunan, Hubei, and Chongqing.

89. Jingliangyou 1212

The two-line indica hybrid rice bred by Yuan Longping High-Tech Agriculture Co. Ltd. was recognized as a super rice variety in 2019. It has an average growth period of 153.7 days when planted in the middle reaches of the Yangtze River for single-cropping mid-season rice cultivation. The comprehensive index for rice blast disease over two years was 2.7 and 2.5, respectively, with the highest rating for panicle blast at level 3, and brown planthopper at level 9. It is moderately resistant to rice blast disease, highly susceptible to brown planthopper, and has strong heat tolerance during the heading stage and moderate cold tolerance. The main rice quality indicators are a head rice rate of 68.4%, a chalky rice rate of 10%, a chalkiness degree of 2.7%, an amylose content of 15.2%, a gel consistency of 63 mm, and a length-to-width ratio of 2.9, meeting the standard of grade 2 in

Cooking Rice Variety Quality for the agricultural industry. When planted in the middle and lower reaches of the Yangtze River for single-cropping mid-season rice cultivation, it has an average growth period of 133.3 days. The comprehensive index for rice blast disease over two years was 3.1 and 3.0, respectively, with the highest rating for panicle blast at level 3, indicating moderate resistance to rice blast disease, susceptibility to bacterial leaf blight, and high susceptibility to brown planthopper. It has moderate heat tolerance during the heading stage. The main rice quality indicators are a head rice rate of 60.8%, a chalky rice rate of 11%, a chalkiness degree of 3.0%, an amylose content of 13.7%, a gel consistency of 73 mm, and a length-to-width ratio of 3.2, meeting the standard of grade 2 in *Cooking Rice Variety Quality* for the agricultural industry. In regional trials conducted in the upper reaches of the Yangtze River for the mid-season late-maturing indica rice group in 2016 and 2017, the average yield per *mu* was 622.51 kg, while in production trials in 2017, the average yield per *mu* was 621.66 kg. In regional trials conducted in the middle and lower reaches of the Yangtze River for the mid-season late-maturing indica rice group in 2016 and 2017, the average yield per *mu* was 612.10 kg, while in production trials in 2017, the average yield per *mu* was 600.24 kg. This variety is suitable for single-cropping mid-season rice planting in the hilly rice areas of Pingba in Sichuan Province, the low- and mid-altitude indica rice areas in Guizhou Province (excluding the Wuling Mountain area), Yunnan Province, the areas with an altitude below 800 m in Chongqing municipality (excluding the Wuling Mountain area), and southern rice area of Shanxi Province. It is also suitable for single-cropping mid-season rice cultivation in the Yangtze River Basin in Hubei Province (excluding the Wuling Mountain area), Hunan Province (excluding the Wuling Mountain area), Jiangxi Province, Anhui Province, Jiangsu Province, the central rice-growing areas in Zhejiang Province, the northern part of Fujian Province, and the southern part of Henan Province.

90. Huazheyou No. 1

The three-line indica hybrid rice bred by China National Rice Research Institute and Zhejiang Wuwangnong Seeds Shareholding Co. Ltd. was recognized as a super rice variety in 2019. It is suitable for cultivating single-cropping mid-season rice in the middle and lower reaches of the Yangtze River, with an average growth period of 136.5 days. The comprehensive index for rice blast disease was 4.4 for two consecutive years, with the highest ranking for stem blast loss at level 9. It is highly susceptible to rice blast disease, bacterial leaf blight, and brown planthopper. It exhibits strong heat tolerance during the heading stage. The main quality indicators include a head rice rate of 63.4%, a chalky rice rate of 14%, a chalkiness degree of 1.7%, an amylose content of 14.7%, a gel consistency of 70 mm, and a length-to-width ratio of 3.0. It meets the third level of the standard in *Cooking Rice Variety Quality* for the agricultural industry. In regional trials conducted in the middle and lower reaches of the Yangtze River in 2016 and 2017 for the mid-season late-maturing indica rice group, the average yield per *mu* was 615.6 kg; while in production trials in 2017, the average yield per *mu* was 609.1 kg. This variety is suitable for single-cropping mid-season rice cultivation in the rice-growing areas of the Yangtze River Basin in Hubei Province (excluding the Wuling Mountain area), Hunan Province (excluding the Wuling Mountain area), Jiangxi Province, Anhui Province, Jiangsu Province, the central rice-growing areas in Zhejiang Province, the northern part of Fujian Province, and the southern part of Henan Province, where the rice blast incidence is low. It is not suitable for cultivation in areas with high rice blast incidences.

91. Wantaiyou 3158

The three-line indica hybrid rice bred by the Rice Research Institute of Guangxi Zhuang Autonomous Region Academy of Agricultural Sciences was recognized as a super rice variety in 2019. It is suitable for early-season rice cultivation in southern Guangxi, with an average growth period of 120.1 days. The comprehensive resistance index for rice blast was 4.8 for two consecutive years, with a maximum panicle blast loss rating of level 5 and bacterial leaf blight rating of 5 to 9. It exhibits moderate susceptibility to rice blast and moderate to high susceptibility to bacterial leaf blight. The main quality indicators include a brown rice rate of 81.1%, a head rice rate of 55.6%, a chalkiness degree of 0.5%, a chalky rice rate of 8%, a length-to-width ratio of 3.4, a transparency level 2, an alkali spreading value of 4.8, a gel consistency of 78 mm, and an amylose content of 13.3%. In regional trials conducted in southern Guangxi for early-season late-maturing rice in 2016 and 2017, the average yield per *mu* was 568.7 kg; while in production trials in 2017, the average yield per *mu* was 535.6 kg. This variety is suitable for early-season rice cultivation in the rice-growing areas of southern Guangxi, while in other rice-growing areas, the appropriate planting season should be selected based on variety trial demonstrations. Attention should be paid to the prevention and control of diseases and pests, such as rice blast.

92. Jingliangyou 1988

The two-line indica hybrid rice bred by Yuan Longping High-Tech Agriculture Co. Ltd. was recognized as a super rice variety in 2020. It is suitable for planting in the middle and lower reaches of the Yangtze River as a single-cropping mid-season rice variety, with an average growth period of 135.8 days. Resistance: The comprehensive resistance index for rice blast disease is at level 2.6, with a maximum panicle blast loss rating of 3, showing moderate resistance to rice blast disease, susceptibility to bacterial leaf blight, high susceptibility to brown planthopper, and strong heat tolerance during heading. The main quality indicators include a head rice rate of 70.1%, a chalkiness degree of 17%, a chalky rice rate of 4.3%, an amylose content of 14.8%, a gel consistency of 71 mm, and a length-to-width ratio of 3.1. For double-cropping late-season rice planting in South China, the variety has an average growth period of 117.4 days. Resistance: The comprehensive resistance index for rice blast in two consecutive years is 3.3 and 3.9, respectively, with a maximum panicle blast loss rating of 3, bacterial leaf blight rating of 7, and brown planthopper rating of 9. It shows moderate resistance to rice blast, susceptibility to bacterial leaf blight, and high susceptibility to brown planthopper. The main quality indicators include a head rice rate of 57.8%, a chalkiness degree of 15%, a chalky rice rate of 3.3%, an amylose content of 15.8%, a gel consistency of 60 mm, and a length-to-width ratio of 3.1. In regional trials conducted in the middle and lower reaches of the Yangtze River for mid-season late-maturing indica rice groups in 2016 and 2017, the average yield per *mu* was 632.97 kg. In production trials in 2017, the average yield per *mu* was 647.27 kg. In regional trials conducted in South China for photosensitive late-season indica rice groups in 2017 and 2018, the average yield per *mu* was 493.89 kg. In production trials in 2018, the average yield per *mu* was 476.73 kg. This variety is suitable for single-cropping mid-season rice planting in the rice-growing areas of Hubei Province (except for the Wuling Mountain area), Hunan Province (except for the Wuling Mountain area), Jiangxi Province, Anhui Province, Jiangsu Province in the Yangtze River Basin, the middle rice-growing areas of Zhejiang Province, the northern rice-growing area of Fujian Province, and the southern rice-growing area of Henan Province. It is also suitable for late-season rice planting in the light incidence areas of bacterial leaf blight in the double-cropping rice areas of Guangdong Province (except for the rice cultivation areas of northern Guangdong), Guangxi, southern Guangxi, Hainan Province, and southern Fujian Province.

93. Jiafengyou 2

The three-line indica hybrid rice bred by Zhejiang Kedefeng Seed Industry Co. Ltd. and Jiaxing Academy of Agricultural Sciences was recognized as a super rice variety in 2020. In the trials conducted for single-cropping hybrid indica rice in Zhejiang Province in 2015 and 2016, the average yield per *mu* was 673.5 kg, and it reached 667.8 kg in the synchronized participation in provincial production trials in 2016. The average growth period in the two-year trial period in the province was 144.7 days. Based on the resistance identification conducted by the Zhejiang Academy of Agricultural Sciences from 2015 to 2016, the maximum rating for panicle blast loss is at level 1, with a comprehensive index of 2.3. The maximum rating for bacterial leaf blight is at level 7, and the maximum rating for brown planthopper is at level 9. As tested by the Rice and Product Quality Supervision and Testing Center of the Ministry of Agriculture and Rural Affairs from 2015 to 2016, the variety shows an average head rice rate of 64.1%, a length-to-width ratio of 2.7, a chalky rice rate of 9%, a chalkiness degree of 0.8%, a transparency level of 2, an alkali spreading value of 6.3, a gel consistency of 78 mm, and an amylose content of 15.1%. The comprehensive evaluation of rice quality indicators for two consecutive years was rated as class 2 by the Food Rice Quality Department. This variety is suitable for single-cropping indica rice cultivation in the rice-growing areas of Zhejiang Province.

94. Huazheyou 71

The three-line indica hybrid rice bred by China National Rice Research Institute and Zhejiang Wuwangnong Seeds Shareholding Co. Ltd. was recognized as a super rice variety in 2020. In the trials conducted for single-cropping hybrid indica rice in Zhejiang Province in 2015 and 2016, the average yield per *mu* was 631.3 kg, and it reached 589.8 kg in the synchronized participation in provincial production trials in 2016. The average growth period in the two-year trial period in the province was 136.9 days. Based on the resistance identification conducted by the Zhejiang Academy of Agricultural Sciences from 2015 to 2016, the maximum rating for panicle blast loss is at level 5, with a comprehensive index of 4.5. The maximum rating for bacterial leaf blight is at level 9, and the maximum rating for brown planthopper is at level 9. As tested by the Rice and Product Quality Supervision and Testing Center of the Ministry of Agriculture and Rural Affairs in 2015–2016, the variety shows an average head rice rate of 62.9%, a length-to-width ratio of 2.9, a chalky rice rate of 18%, a chalkiness degree of 2.9%, a transparency level of 2, an alkali spreading value of 5.5, a gel consistency of 86 mm, and an amylose content of 15.3%. The comprehensive evaluation of rice quality indicators for two consecutive years was rated as class 3 by the Food Rice Quality Department. This variety is suitable for single-cropping indica rice cultivation in Zhejiang Province and can also be planted as early- or mid-season rice in the central and northern regions of Guangxi, as mid-season rice in the alpine region, and as early- or late-season rice in the southern rice-growing areas of Guangxi.

95. Funongyou 676

The three-line indica hybrid rice bred by the Rice Research Institute of Fujian Academy of Agricultural Sciences and Fujian Hefeng Seed Industry Co. Ltd. was recognized as a super rice variety in 2020. The average growth period for this variety is 144.0 days. The comprehensive evaluation of resistance to rice blast from two-year trials shows it to be moderately resistant. In terms of rice quality testing results, the brown rice rate is 79.3%, the head rice rate is 62.1%, the chalkiness degree is 1.5%, the transparency level is 1, the alkali spreading value is 5.4, the gel consistency is 86 mm, and the amylose content is 16.0%. In the two-year trials for late-season indica rice varieties conducted in 2016 and 2017, the average yield per *mu* was 648.81 kg. In the mid-season rice production test conducted in Fujian Province in 2017, the average yield per *mu* was 597.34 kg. This variety is suitable for mid-season rice cultivation in Fujian Province.

96. Longfengyou 826

The three-line indica hybrid rice bred by the Rice Research Institute of Guangxi Academy of Agricultural Sciences was recognized as a super rice variety in 2020. The average growth period for this variety was 115.4 days. The comprehensive evaluation of resistance to rice blast disease from two-year trials shows it has an index of 3.0 and 5.8, with a maximum panicle blast loss rate of 7, and a 9 rating for bacterial leaf blight. It was moderately resistant to rice blast and highly susceptible to bacterial leaf blight. In terms of rice quality testing results, the brown rice rate is 81.5%, the head rice rate is 62.7%, the length-to-width ratio is 3.0, the chalky rice rate is 13%, the chalkiness degree is 1.5%, the transparency level is 1, the alkali spreading value is 6.9, the gel consistency was 80 mm, and the amylose content is 15.6%, meeting the standard of class 2 in the *Cooking Rice Variety Quality* from the Ministry of Agriculture and Rural Affairs. In 2015 and 2016, the average yield per *mu* of the late-season indica group in Guangxi reached 521.4 kg for two consecutive years. In the production test conducted in 2016, the average yield per *mu* was 496.4 kg. This variety is suitable for late-season rice cultivation in areas that are suitable for planting photosensitive varieties in the southern and central regions of Guangxi.

97. Taiyou 871

The three-line indica hybrid rice bred by the School of Agricultural Sciences of Jiangxi Agricultural University and Rice Research Institute of Guangdong Academy of Agricultural Sciences was recognized as a super rice variety in 2020. The average growth period for this variety is 121.4 days. It has a brown rice rate of 81.7%, a milled rice rate of 70.0%, a head rice rate of 64.5%, a grain length of 7.6 mm, a length-to-width ratio of 4.0, a chalky rice rate of 11%, a chalkiness degree of 2.7%, an amylose content of 16.3%, and a gel consistency of 60 mm. The rice quality reached the national standard of grade 2 high-quality rice according to *High Quality Paddy*. According to the natural induction identification for resistance to rice blast disease, it has a rating of 9 for stem blast, indicating high susceptibility to rice blast, and a comprehensive resistance index of 4.6. In the provincial trials conducted in Jiangxi in 2014 and 2015, the average yield per *mu* was 594.88 kg. This variety is suitable for cultivation in Jiangxi Province.

98. Jingyou Huazhen

The three-line indica hybrid rice bred by Sichuan Ludan Zhicheng Seed Industry Co. Ltd. and the Rice and Sorghum Research Institute of Sichuan Academy of Agricultural Sciences was recognized as a super rice variety in 2020. It is suitable for single-cropping mid-season cultivation in the upper reaches of the Yangtze River, with an average growth period of 158.4 days. The comprehensive resistance index for rice blast was 5.1 and 5.8 in two consecutive years, with the highest rating for panicle blast damage being level 7. It is susceptible to rice blast and shows high susceptibility to brown planthopper with a rating of 9. The main indicators of rice quality include a head rice rate of 60.8%, a length-to-width ratio of 3.2, a chalky rice rate of 14%, a chalkiness degree of 1.2%, a gel consistency of 88 mm, and an amylose content of 18.3%. It meets the national standard of grade 2 high-quality rice according to *High Quality Paddy*. In 2014 and 2015, its average yield per *mu* in the trials for the mid-season late-maturing indica rice group in the upper reaches of the Yangtze River reached 624.3 kg for two consecutive years. In the production trial conducted in 2016, the average yield per *mu* was 605.4 kg. This variety is suitable for single-cropping mid-season cultivation of in the hilly rice areas in Pingba of Sichuan Province, except for the Wuling Mountain region, and in the low- and mid-altitude indica rice areas of Guizhou Province (excluding the Wuling Mountain region), Yunnan Province, and areas with an altitude below 800 m in Chongqing Municipality (excluding the Wuling Mountain region), and in the southern rice areas of Shanxi Province.

99. Longfengyou 826

The three-line indica hybrid rice bred by the Rice Research Institute of Guangxi Academy of Agricultural Sciences was recognized as a super rice variety in 2020. The average growth period for this variety is 115.4 days. The comprehensive evaluation of resistance to rice blast disease from two-year trials shows it to have an index of 3.0 and 5.8, with a maximum panicle blast loss rate of 7, and a 9 rating for bacterial leaf blight. It is moderately resistant to rice blast and highly susceptible to bacterial leaf blight. In terms of rice quality testing results, the brown rice rate is 81.5%, the head rice rate is 62.7%, the length-to-width ratio is 3.0, the chalky rice rate is 13%, the chalkiness degree is 1.5%, the transparency level is 1, the alkali spreading value is 6.9, the gel consistency is 80 mm, and the amylose content was 15.6%, meeting the standard of class 2 in the *Cooking Rice Variety Quality* from the Ministry of Agriculture and Rural Affairs. In 2015 and 2016, the average yield per *mu* in the trial for the late-season indica group in Guangxi reached 521.4 kg for two consecutive years. In the production test conducted in 2016, the average yield per *mu* was 496.4 kg. This variety is suitable for late-season rice cultivation in areas that are suitable for planting photosensitive varieties in the southern and central regions of Guangxi.

100. Yongyou 7850

The three-line hybrid rice of indica-japonica hybridization bred by Ningbo Seed Co. Ltd. was recognized as a super rice variety in 2020. It is primarily cultivated as a single-cropping mid-season rice in the upper reaches of the Yangtze River, with an average growth period of 158.4 days. The comprehensive resistance index for rice blast was 5.1 and 5.8 in two consecutive years, with the highest rating for panicle blast damage being level 7. It was susceptible to rice blast and also showed high susceptibility to brown planthopper, with a rating of 9. The main indicators of rice quality include a head rice rate of 60.8%, a length-to-width ratio of 3.2, a chalky rice rate of 14%, a chalkiness degree of 1.2%, a gel consistency of 88 mm, and an amylose content of 18.3%. It meets the national standard of grade 2 high-quality rice according to *High Quality Paddy*. In 2014 and 2015, the average yield per *mu* in the trial for the mid-season late-maturing indica rice group in the upper reaches of the Yangtze River reached 624.3 kg for two consecutive years. In the production trial conducted in 2016, the average yield per *mu* was 605.4 kg. This variety is suitable for single-cropping mid-season rice cultivation in the hilly rice areas of Pingba in Sichuan Province, and in the low- and mid-altitude indica rice areas of Guizhou Province (excluding the Wuling Mountain region), Yunnan Province, and areas with an altitude below 800 m in Chongqing Municipality (excluding the Wuling Mountain region), and in the southern rice areas of Shanxi Province. It is also suitable for early- or mid-season rice cultivation in the central and northern regions of Guangxi, and for mid-season rice cultivation in the alpine region.

CHAPTER 3

Cultivation Techniques for Strong Seedling of Super Hybrid Rice

I. The Roles and Standards of Strong Seedlings

With a history of over 1,800 years, rice seedling transplantation remains one of the most effective ways of cultivating rice in China. Cultivating strong seedlings lays the foundation for high-yielding rice. Just as the saying goes, "Good seedlings are half of success." Nurturing sufficient high-quality strong seedlings is the only way to take the initiative at the very beginning, giving full play to seedlings' advantages and striving for overall high yields.

1. The Roles of strong seedlings

1) *Substitute tillers for seedlings to reduce seed consumption.* The tillering of strong seedlings follows the pattern of leaf-tiller elongation. Seedling fields boast many tillers, and the tillers with more than two leaves and a central bud can further tiller and develop into panicles just like mother stems. By substitut-

ing tillers for seedlings, the number of seeds required per *mu* reduces, resulting in a more cost-friendly effect.

2) *High physiological activity is conducive to early development.* With high physiological activity, strong seedlings exhibit strong stress resistance, low leaf-withering rate, vigorous rooting ability, rapid regreening, and early tillering once transplanted to the field. These characteristics are helpful for early growth and early maturation and maximize the utilization of the growing season.
3) *Facilitate the formation of high-yielding populations.* Cultivating strong seedlings lays the foundation for establishing a high-yielding population. As strong seedlings tiller extensively in the seedbed and grow rapidly after transplantation, they produce numerous tillers at low nodes in the early stage, ensuring adequate ears and benefiting the formation of large panicles. A higher number of grains per panicle benefits the production of photosynthetic products and their transportation to the panicle, increasing dry matter production and economic coefficient and forming a high-yielding population featuring "early development, mid-term stability, and late-stage resilience."

The standards for strong seedlings vary depending on the seedling cultivation method, seedling age, and variety type.

2. The indicators of strong seedlings

1) *Morphological indicators.* Here are four requirements for strong seedlings in morphology. First, the leaves should be large, vigorous, not soft or drooping, with short leaf sheaths and stout and flat plant stems (pseudostem), known as broad-leaf seedlings. Second, the leaves should be dark green, with neither too thick nor too light color, free from insect damage or disease spots, with more green leaves and fewer yellowed leaves. Third, a well-developed root system should be with many short white roots. Fourth, seedlings should grow uniformly, with few lanky or weak seedlings, exhibiting resilience in plant body and strong growth momentum.
2) *Physiological indicators.* First, strong seedlings should show characteristics of strong photosynthetic capacity, abundant nutrient reserves in the body, high fullness ratio (dry weight/seedling height), significant dry weight per hundred plants, generally above five g for small seedlings and above seven g for large seedlings. Second, they are expected to have a balanced carbon-to-nitrogen ratio (C/N), with high levels of carbohydrates and nitrogen compounds in the plant to prevent aging from excessive carbon content or weakness from excessive nitrogen content. The C/N ratio should be around 14 for large seedlings and around 10 for medium-sized seedlings. Third, strong seedlings have higher bound water content and lower free water content, which is conducive to post-transplantation water balance, enhancing seedling stress resistance, with the bound water content of leaves above 30%.
3) *Post-planting growth indicators.* First, the seedlings should have strong post-plant root regeneration strength, with thick and flat stem bases (pseudostems), numerous root primordia,

high absolute content of carbohydrates in the body, rapid cell proliferation rate, and strong root regeneration ability. Second, they should show a low rate of plant damage, with many short white roots in strong seedlings, rapid recovery of root absorption function, abundant nutrient reserves in the seedling body, high bound water content, and strong resistance to adverse environmental conditions such as intense heat or cold and rainy weather and quick re-greening and early tillering after transplantation.

II. Seed Treatment and Seed Soaking for Germination

1. Seed drying

Sun exposure enhances seed permeability and water absorption capacity, increases enzyme activity, and improves seed germination rate and vigor, resulting in a 2% to 5% increase in germination rate. Seeds are usually sun-dried on five to seven sunny days before sowing, with a light exposure for one day.

2. Seed selection

Seed selection is conducive to removing blighted grains, diseased grains, weed seeds, and impurities to select full, neat, and clean seeds for cultivating strong seedlings and reducing weed growth. Commercial seeds sold by seed companies nowadays have been sorted by seed cleaners. Although their purity, cleanliness, and germination rate have met certain standards, it is still necessary to wash and select seeds. Due to the particularities of hybrid seeds, it is not recommended to select seeds with salt water. After soaking in clean water for about 12 hours, separate the plump seeds that sink to the bottom of the water from the floating blighted grains to soak, disinfect, and accelerate germination; subsequently, sow and manage separately to save hybrid rice seeds.

3. Seed disinfection

Effective seed disinfection is vital in preventing the transmission of various diseases and pests in rice, such as rice blast, bakanae disease, bacterial leaf blight, and rice white tip.

4. Seed soaking

Rice seeds need to absorb water equal to about 25% of their own weight to germinate. Therefore, to promote uniform germination and seedling emergence, it is essential to ensure that the seeds are soaked for an adequate amount of time to absorb sufficient water. The soaking time of seeds is closely related to factors such as temperature, variety, and seedling cultivation methods. In high temperatures, the soaking time is shorter, while in low temperatures, it is longer. For dry seedling cultivation, the soaking time can be longer, while for wet seedling cultivation, it can be shorter. Hybrid rice seeds have a high respiration rate during soaking, resulting in increased carbon dioxide production. It is important to regularly change the water, use flowing water for soaking, or adopt a "three up, three down" method. The characteristic of seeds fully absorbing water is indicated by the semi-translucent husk, distinct white belly, and swollen embryo. Generally, seeds need to soak for 72 hours at 10°C, 48 hours at 20°C, and only 24 hours at 30°C to reach the required moisture level for germination.

1) *Warm water soaking method.* This is an effective method to prevent and control rice white tip disease. First, soak the seeds in clean water for 24 hours, then preheat them in water at 45°C to 47°C for five minutes. Then, immerse them in water at 50°C to 52°C for ten minutes to kill nematodes. Subsequently, transfer the seeds to cold water for further soaking until reaching the germination requirement. This method can eliminate pathogens of diseases such as rice blast and bakanae disease.
2) *Chlorine disinfection soaking method.* Pre-soak the seeds for 12 to 24 hours, then immerse them in a solution of strong chlorine bleach diluted 250 to 300 times for 12 hours. After rinsing with clean water, continue soaking or germinating the seeds. This method has a positive effect on preventing and controlling various diseases.
3) *Zhongshengmycin soaking method.* First, soak the seeds intermittently in water, then immerse them in a solution of 3% Zhongshengmycin wettable powder diluted 300 to 400 times for 10 to 12 hours. Wash the seeds and initiate germination after soaking.
4) *Limewater soaking method.* It is advisable to soak the seeds in a 1% clear limewater solution. During soaking, it is important to avoid disrupting the lime film on the water surface. Additionally, make sure the water depth is at least three cm above the seeds. This method is effective in preventing and controlling various diseases.

5. Germination acceleration

Germination acceleration ensures uniform seedling emergence and prevents seed from rotting. This is crucial in the early rice sowing period when the temperature is low. Generally, seeds are preheated in water at 50°C for five to ten minutes and then drained, sealed, and insulated at the temperature of 30°C to 35°C. Once the sprouts appear and the weather is ideal, the seeds can be sown. In case of unfavorable weather conditions, promote root growth at 25°C to 30°C, then spread sprouts to allow germination, maintaining a certain temperature and humidity. The sprout length should not exceed half the length of the grain, and the roots should not be longer than the grain. Special attention should be paid to prevent buds from being injured by high temperatures. During the soaking of mid- and late-season rice varieties, both the water temperature and air temperature are relatively high, with the daily average temperature exceeding 25°C. In this case, seeds can be soaked while allowing them to naturally germinate by adopting the "three up, three down" method at room temperature, which is simple, practical, and effective.

III. Seedling-Breeding Techniques of Super Hybrid Rice

There are two main methods for seedling breeding of super rice. One is to adopt a semi-dry seedling-breeding technique with plastic trays to cultivate nursing seedlings or small seedlings that meet the requirements of the intensive cultivation system. The other is to adopt the conventional sparse seeding method in a wet seedling field, cultivating medium seedlings or large seedlings for pulling and transplanting.

1. Plastic-tray seedling-breeding method for cultivating nursing seedlings or small seedlings suitable for the system of rice intensification

The plastic-tray seedlings provide multiple benefits, such as seed and nursery field reduction, conservation of fertilizer and water, decreased reliance on agricultural films, labor-saving, and increased productivity. Besides being especially suited to seedling throwing, they are ideal places for cultivating nursing seedlings or small seedlings with shallow soil depth. The steps are as follows:

1) *Prepare nutrient soil.* Choose fertile vegetable garden soil mixed with well-rotted organic fertilizer, or weed-free fertilizer compost as nutrient soil, or incorporate 1.5% of a seedling-strengthening agent based on the total weight of the soil to ensure even and comprehensive nutrient distribution. Each plastic tray requires 1.5 kg of dry fine soil or 2.0 kg of compost. Generally, the effective nutrients in dry soil (or red-yellow soil) are insufficient. Directly adopting it as nutrient soil may result in weak and stunted seedlings, so it is necessary to supplement with quick-acting nutrients such as phosphorus, potassium, and urea, which can be mixed at 0.5%, 0.2%, and 0.1% of the weight of the soil respectively as a base fertilizer, and compound fertilizers can be used up to 0.3%. Solid fertilizers have limited mobility in the soil; they should be dissolved and diluted in water before being evenly sprayed and mixed into the soil to prevent seedling damage caused by excessive fertilization. Under low-temperature conditions, every 100 kg of soil should be disinfected with 10 g of hymexazol dissolved in 15 kg of water.

2) *Tidy up the dry seedling bed.* Select leeward and sun-exposed vegetable plots with fertile soil, good water permeability, and convenient drainage and irrigation; or use paddy fields that have been plowed and baked before winter as high-standard dry seedbeds (or semi-dry seedbeds). Prepare seedling beds equal to 1/40 of the total area. Dig furrows at intervals of 1.8 m, with a width of 1.45 m, a furrow width of 0.35 m, and a depth of 0.2 m. The length of the furrows depends on the shape of the field, with surrounding ditches. After forming the ridges, break up the surface soil layer, water (or irrigate) and level it out, spread a layer of slurry, apply a light topdressing, and level and press down the furrow beds.
3) *Sow and place trays.* The plastic trays for seedling cultivation measuring 60 cm × 33 cm × 1.8 cm are on trend now, with 353 holes each, and about 35 trays are recommended per *mu*. There are two different methods, directly placing the trays on seedling beds and then sowing or sowing in the trays and then transferring them to the seedling nursery. The latter one is more common in practice, and its specific operations are as follows: Put a floppy tray on the roadside near the seedling field or on an open shed ground and equip each tray hole with nutrient soil of no more than 2/3 of the hole depth (dry soil must be fully watered). Sow one to two grains of rice seeds (rice seeds refer to dehulled grains, with root buds no longer than 1.5 mm) in each hole, with a seeding rate of approximately 0.5 kg/*mu*. Super rice seeders can be adopted to improve efficiency and reduce labor intensity. After sowing, cover the surface of the tray with nutrient soil (or diluted mud) to expose no seeds, wipe off the soil higher than the tray, and spray the seedling trays with water. Then, arrange the seedling trays, ensuring that each seedling tray is closely connected, and the bottom of the seedling trays is in close contact with the slurry layer on the bed surface to prevent dehydration and seedling death caused by moisture loss.
4) *Cover with plastic film for seedling protection.* Setting up a support frame and covering it with agricultural film after sowing contributes to moisture retention, cold protection, rain protection, and rodent and bird prevention. However, attention should be paid to preventing the emergence of leggy seedlings and scorching seedlings due to high temperatures. Timely removal of the film for cooling down and hardening the seedlings is necessary.
5) *Manage the seedling bed.* The key points of management are "controlling seedlings with water, promoting roots with fertilizer, and promoting tillers with roots." It is essential to maintain a moist bed surface before the emergence of all seedlings. Apply organic liquid manure on the one-leaf-one-bud stage. At the stage of 1.5 leaves, irrigate (a sprinkler or similar equipment can be used) with a 1% urea solution as the weaning fertilizer, at a rate of 1.5 kg/*mu* (if the seedling-strengthening agent does not contain paclobutrazol or has an inconspicuous dwarfing effect, after the foliage dries, evenly spray once with a solution of 150 g of 15% paclobutrazol wettable powder dissolved in 75–100 kg of water). On rainy days, it is important to control water, prevent waterlogging, keep the leaves from curling in the morning and evening, and prevent the surface from turning white or excessive accumulation of water. If there are sunny days with high temperatures after sowing, promptly remove the plastic film to cool down and irrigate the surrounding ditches to keep the bed surface moist. During the 1.5 to 2.1 leaf stage, it depends on the situation to remove the plastic film. Lightly applying a starter fertilizer (adopt 2.5 kg of urea per *mu*, mixed with 300–400 kg of water

for irrigating and sprinkling over the seedlings) two days before transplanting is conducive to tillering.

2. Conventional seedling breeding methods for cultivating large seedlings (or medium seedlings)

Wet seedling cultivation is the most adopted method for seedling cultivation. It falls between water seedling raising and dry seedling raising and is commonly referred to as semi-dry seedling cultivation. The distinctive features of it involve plowing and leveling the field in a dry condition, then watering and smoothing it out, which creates a seedbed that is compact on top and loose underneath, ensuring good ventilation. From sowing to root establishment, the field is kept moist and aerated to facilitate root growth. Combined with field drainage, frequent shallow water irrigation is adopted from root establishment to the three-leaf stage. After that, shallow water is irrigated along the ridges, which is different from traditional water-based seedling cultivation. The following are the three advantages: First, its soil aeration is better than water-based seedling cultivation, which promotes root growth and greatly improves the seedling rate. Second, it enhances seedling quality and promotes early and vigorous growth after transplantation. Third, it boosts the resistance of seedlings and reduces the emergence of putrid seedlings.

When cultivating single-cropping mid-season rice or single-cropping late-season rice, the high temperatures lead to the rapid growth of seedlings, thus it is crucial to focus on the following aspects of nursery field management:

1) *Strengthen water regulation.* From sowing to the one-leaf-one-bud stage, keep the ridge surface dry but irrigate the furrows to prevent heat-induced germination. To prevent scattering of the seeds on heavily rainy days, the ridge surface can be temporarily filled with water and drained immediately after the rain. Keep water in the furrows from the one-leaf-one-bud stage to the two-leaf-one-bud stage to avoid surface cracking and refrain from watering on the surface. If cracking occurs, irrigate with the running water technique. After the three-leaf stage, irrigate with shallow water on the surface to promote tillering. In high-temperature days, irrigate in the daytime and drain at night to cool down if possible.
2) *Topdressing in time.* Apply the weaning fertilizer once more during the one-leaf-one-bud stage, the relay fertilizer during the four-to-five-leaf stage, and the transplanting fertilizer three to five days before transplanting. The amount of fertilizer applied each time should be moderate to avoid excessive growth. It is advisable to apply three to four kg of urea and potassium chloride per *mu* of seedling field.
3) *Growth regulation and tiller promotion.* For seedlings that have not been treated with uniconazole, spray 200 g of 15% paclobutrazol mixed with 100 kg of water per *mu* during the one-leaf-one-bud stage to control seedling growth and promote tillering. Seedlings that have been treated with uniconazole and are approaching 30 days of age can also be sprayed again during the three-leaf-one-bud stage with paclobutrazol.

4) *Even spacing of seedlings.* During the two-leaf-one-bud stage, it is essential to perform the thinning operation once to ensure uniform growth of seedlings and facilitate consistent individual development by evenly spacing seedlings.
5) *Prevention and control of diseases and pests.* The main targets for disease and pest control in mid-season and late-season rice seedling fields include rice thrips, rice planthopper, rice leafhoppers, Cnaphalocrocis medinalis, Chilo suppressalis, and rice blast. Some rice-growing areas may also suffer from gall midge and hydrellia griseola. It is important to adopt pesticides and spray them in time. For effective control of rice thrips, pesticides should be applied every four to five days, and the pre-transplantation drug should be applied two days before transplanting, saving the cost of field control measures.

IV. Mechanical Rice Direct-Seeding Technology

1. Field preparation

1) *Apply enough base fertilizer.* Before plowing wheat stubble rice, evenly distribute 50 kg of cake fertilizer, 1,000 to 1,500 kg of barnyard manure, and 100 kg of organic bio-fertilizer per *mu*. For green manure rice (with a biomass of 1,500 kg), apply 25 kg of ammonium bicarbonate and compound fertilizer (15% nitrogen, 15% phosphorus, and 15% potassium) respectively per *mu*.
2) *Field preparation and quality.* For fallow fields, plow to a depth of 15 to 20 cm, perform water rotation tillage, and level the field three to five days before sowing. For wheat stubble fields, perform water rotation tillage to a depth of 10 to 15 cm after wheat harvest, and level the field two to four days before sowing. Special attention should be paid to achieving a flat surface during field preparation, ensuring that the height difference of the whole field does not exceed 3.3 cm, the field is flat and deprived of obvious high ridges or low areas, and the internal and external ditches should be coordinated in a timely manner. In wet direct-seeding paddies, precautions should be taken to avoid excessively muddy conditions, which may lead to seed sealed. Sow the seeds when the mud is firm but with a moderate hardness, ensuring that the seeding process does not cause mud surges and the seeds are lightly covered by a thin layer of mud slurry. Adjust the softness and hardness of the mud by controlling the drainage time. Generally speaking, drain the water before sowing, and sow the seeds when most of the water has been released. If the mud is excessively muddy, drain the water the other day, and sow the seeds when the mud reaches a suitable firmness. To prevent sealed seeds and seedling shortage caused by mud surges, it is important to avoid sowing in excessively muddy conditions.

2. Mechanical rice direct-seeding technology

1) *Sowing period.* Timely sowing is one of the key cultivation measures for achieving high yields. Different regions should arrange their sowing periods based on local conditions. In Hunan, the single-cropping rice is usually sown around May 15, but no later than May 25.
2) *Seeding quantity.* Excessive basic seedlings are one of the main problems in mechanical direct-seeding, which results in significant imbalances between regions and fields. The number of basic seedlings is closely related to the seeding quantity and seedling rate. From the perspective of production techniques, wide-row sowing is advocated, with 8 rows every 2 m and 30 to 33 seeds per m in the sowing row. If the sowing width is 2 m with 10 rows, 24 to 26 seeds will be sown per m of the sowing row. The sowing density of conventional rice is doubled based on hybrid rice; for example, if the sowing width is 2 m with 10 rows, the sowing row will be 48 to 52 seeds per m. Pay close attention to whether there are straws dragged behind the carriage. If these straws drag off the sown grain seeds, stop sowing immediately and step the straws dragged behind into the mud. Replant them manually in places that cannot be sown by machine at the four corners of the field and avoid sowing too dense.
3) *Seeding depth.* Maintaining the proper seeding depth is key to ensuring complete, uniform, and strong seedlings. The seeding depth is determined by the seed's topsoil ability, cultivation system, soil type, and climate conditions. For drill seeding, the general seeding depth is one to three cm, with a maximum depth of three cm. For upland rice fields where the emergence of seedlings relies on soil moisture before sowing, a seeding depth of two to three cm is suitable. For the upland rice fields where the seeds emerge with water after sowing, it is necessary to sow shallowly with a depth of one to two cm.

3. Fertilizer and water management

1) *Strengthen post-sowing management.* First, open the ditch immediately after sowing and connect it with the sowing tank to discharge the accumulated water in the field. Second, fill vacancies timely for uniform seedling growth.
2) *Regulate the fertilizer and water reasonably and adjust the population structure.* First, apply reasonable fertilizer regulation with nitrogen reduction and phosphorus and potassium increase. Fertilizer management should continue adhering to the stable use of basal-tillering fertilizer, reduce long coarse fertilizer, and apply panicle fertilizer appropriately. According to the experiment and production practice, the nitrogen application rate of super rice per *mu* is 12 to 14 kg (equivalent to 70 to 85 kg of ammonium bicarbonate), among which basal-tillering fertilizer accounts for about 80%. In mid-to-late July, the dosage of nitrogen fertilizer should be controlled to control the growth of high tillers and the lengths of the fifth-to-last and the fourth-to-last leaf. At the end of July and the beginning of August, based on putting aside good fields, seedlings should be supplemented with panicle fertilizer depending

upon their specific condition. Each *mu* should use 10 kg of 45% high-efficiency compound fertilizer or 15 kg of 25% compound fertilizer, and 10 kg of potassium chloride should be added to make the potassium-nitrogen ratio reach 0.5. Controlling nitrogen fertilization and increasing potassium fertilization are important measures to improve the yield and quality of hybrid rice. Second, the ratio of nitrogen, phosphorus, and potassium fertilizers in hybrid rice is 1:0.7:1.1. In the application of base fertilizer, advocate organic fertilizer and reduce inorganic nitrogen fertilizer. In the application of panicle fertilizer, advocate compound fertilizer, reduce nitrogen fertilizer, and actively promote organic liquid fertilizer. Third, coordinate the water and slurry, extend the control period, and ensure proper water and slurry management. During the direct-seeding stage, the focus should be on keeping the soil moist. After the two-leaf-one-bud stage of rice transplanting, establish a shallow water layer two to three days after transplanting, with primary irrigation during the tillering stage focused on shallow water irrigation. Once the expected number of panicles is reached, drain the water step by step and gradually increase the intensity. Strictly control the peak seedling population to maintain an appropriate plant structure and good ventilation and light conditions, thereby increasing the tillering and heading rate. In the middle stage, implement shallow wet irrigation, transitioning to intermittent dry and wet conditions in the later stages and not cutting off the water supply too early.

V. Application of Plant Growth Regulators in Seedling Breeding of Super Hybrid Rice

Plant growth regulators are effective in cultivating strong seeds. At present, the most adopted products are uniconazole, paclobutrazol, and other products containing chemical control agents. Typically, applying one type is sufficient to control elongation and promote tillering. If necessary, such as in the case of seedlings being too mature, two types can be adopted to regulate seedling growth. However, this may not be conducive to realizing the full potential of super-high yields.

1. Uniconazole

Uniconazole is a triazole-containing plant growth retardant that exhibits higher performance and lower residue compared to paclobutrazol. In seedling cultivation, it facilitates dwarfing and tillering, increases chlorophyll content, stimulates root growth, and enhances seedling resistance. Soaking the seeds in a uniconazole solution at a concentration of 60–120 mg/kg for over 12 hours is recommended; applying a single spray during the one-leaf-one-bud to the three-leaf-one-bud stage is also recommended. When spraying, ensure that there is no standing water on the seedbed.

2. Paclobutrazol

Paclobutrazol is a triazole-containing plant growth retardant with similar efficacy and mechanism of action to uniconazole. However, it has a longer residual effect period, a lower safety, and is more cost-effective compared with uniconazole. It is recommended to apply a concentration of 250–300 mg/kg of paclobutrazol solution once during the one-leaf-one-bud to the three-leaf-one-bud stage. Generally, it is not used for seed soaking; and for older seedlings, paclobutrazol can be applied after soaking with uniconazole.

3. Other growth regulators

To simplify seedling-raising procedures, uniconazole or paclobutrazol are commonly used as raw materials in production. They are often formulated with other nutrients, fungicides, and insecticides to produce specialized seedling-raising products like seedling-strengthening agents, seedling fertilizers, seed-coating agents, and seed-dressing agents, which significantly saves costs. Specific operations should be conducted according to the requirements of each product.

It is advisable to choose seed-coating agents with good film-forming technology. Simply mix the seed coating agent with the seeds thoroughly before soaking, then allow it to dry in the air. Once a film is formed, the seeds can be germinated through soaking.

Seedling-strengthening agents and seedling fertilizers are both designed to simplify the seedling-raising process. They are a type of specialized fertilizers for rice seedling raising that is formulated with chemical regulators, nutrients, insecticides, and fungicides on the basis of the climatic conditions during the early and late rice seedling stages. Also, they have been proven to be highly effective in rice seedling raising, particularly when applied to dry-bed seedling raising and plastic-tray seedling raising. The application of such specialized fertilizers in seedling raising offers the following advantages. First, it simplifies the process by completing fertilization, controlling diseases and pests, and regulating growth in a single application. As these fertilizers have multiple functions, there is no need for additional fertilization, spraying pesticides, or applying growth regulators in the seedling raising period. Second, it helps to cultivate strong seedlings. These products can control excessive elongation of seedlings, promote tillering, increase leaf emergence rate, enhance stem thickness, and improve chlorophyll content in leaves. The quality of comprehensive seedlings raised with these specialized fertilizers is obviously superior to conventional fertilizers. Moreover, they are easy to apply, requiring only one application as base fertilizer, which avoids potential decreases in seedling quality due to factors such as weather conditions and labor constraints that may hinder the implementation of fertilization and pesticide spraying.

The application methods are as follows: for plastic-tray seedling raising, the recommended dosage from the instructions (most manufacturers produce packages equal to the dosage for one *mu* of paddy field) is divided into two equal portions. One portion is mixed evenly with 100 g of dry fine soil per tray and spread uniformly on the seedbed, while the other portion is mixed with paste and directly loaded into the tray. After leveling, the seeds are sown in accordance with the requirements of seedling breeding in plastic trays. For dry-bed seedling breeding, the recommended amount is mixed with one kg of dry fine soil and spread on the seedbed, then evenly incorporated into a two-cm layer of soil using a rake or mixed uniformly and directly laid on the seedbed before watering, sowing, and covering the seeds, following the requirements of dry-bed seedling breeding. Similarly, for wet-bed seedling breeding, the recommended amount per m^2 of seedbed is mixed with one kg of dry fine soil, spread evenly on the seedbed, and then incorporated into a two-cm layer of soil using a rake (or a uniform layer of one to two cm thick paste is laid on the seedbed), followed by sowing and management according to the requirements for wet-bed seedling breeding.

VI. Machine-Transplanted Rice Seedling Cultivation

The key to successful and high-yielding machine-transplanted rice seedling cultivation lies in the density of planting and the high standards compared with traditional conventional seedling raising methods. It is necessary to cultivate standard seedlings that meet the requirements of machine-transplanted seedlings which include uniform emergence, well-developed root systems, strong root binding force, robust individual seedlings, thick stems and roots, and free from disease and pest damage. Transplanting can be carried out when the seedlings reach the three-leaf-one-bud to the four-leaf-one-bud stage with a height of 15 to 20 cm at the age of 20 to 35 days.

1) *Site selection.* Choose a site that is sheltered from the wind, without excessive shading, and has flat terrain without slopes. This will facilitate operations and ensure convenient daily management.
2) *Seedling preparation.* First, prepare materials, including hollow bricks, small bricks, tobacco substrate, sawdust, organic fertilizers, seedling trays, seeds, bamboo strips, plastic film, seed film, shade net, and iron wire. Second, prepare the seedbed. Adopt hollow bricks to enclose a seedling bed measuring 10 m wide. the seedling bed with a two-m-wide plastic film, with the film raised 20 cm on all sides to form a film pool. Spread a layer of 2 cm of sawdust on the surface of the seedling bed. The film pool helps save water and facilitates drainage, and sawdust can absorb water and retain moisture, keeping the seedling trays moist. Third, prepare the seedling-raising substrate. Mix tobacco substrate and organic fertilizers in a ratio of 2:1 before sowing, prepare the seedling-raising substrate, and the seedling-raising substrate should be moist but not sticky when touched. Fourth, process the seeds. Rinse the empty husks from the seeds with clean water and soak them in a diluted solution of prochloraz for 24 hours (to prevent bakanae disease). At last, remove the seeds and drain the excess moisture.
3) *Sow on seedbeds.* Fill the seedling tray with a 2-cm-thick seedling-raising substrate (2.5 cm for a soft tray), ensuring it is evenly leveled. Adopt seeders for sowing (659 seeds per tray for manual seeding). Cover the seeds with seedling-raising substrate after sowing, making sure they are not exposed. Place the filled seedling trays in the seedbeds covered with sawdust,

ensuring five trays each a row and the trays are arranged on one side without any gaps in the middle. For the other side, leave a 5-cm gap for convenient water irrigation and smooth flow.

4) *Water management.* Adopt water hoses to slowly irrigate the seedbed, ensuring the water surface does not reach the seedling trays. Stop watering when the water has penetrated the seed trays. Pour the water from the other side after irrigation and spray with a 500-fold diluted solution of hymexazol (to prevent drooping disease).
5) *Mulching.* Adopt bamboo strips to create arches, inserting them directly into the hollow brick holes, and ensuring the height of the arch should be no less than 50 cm. Cover the arches with perforated mulch film and secure the film in place with small bricks pressed from the sides.
6) *Seedling stage management.* First, water the trays every three to four days to keep the seedling trays moist, ensuring there is no water on the seedbeds. Once the seedlings emerge uniformly after about five days, remove the seedling covers and cover them with a shade net. Ten days later, partially open the sides of the shade net and fasten them with iron wire at a height of 20 cm from the bamboo strips. Fully remove the shade net five days before transplanting and apply insecticide once. Second, uncover or cover the film according to temperature conditions, with the greenhouse temperature maintained at 14°C to 28°C. If the temperature is too high at noon, both ends should be opened for ventilation. In case of low temperatures, do not uncover or change the film. Changing the film after uncovering should be done in the evening after sunset when there is no strong wind.

CHAPTER 4

Fertilization Techniques for Super-High Yield of Super Hybrid Rice

I. Current Situation and Characteristics of Paddy Field Fertility in South China

Indica hybrid rice is mainly distributed in 13 southern provinces (municipalities and autonomous regions), with super indica hybrid rice widely cultivated in provinces (or autonomous regions) such as Hunan, Jiangxi, Sichuan, Fujian, Anhui, Jiangsu, Zhejiang, Guangxi, and Guangdong. Field fertility refers to the soil's ability to provide mineral nutrients in a stable, even, sufficient, and appropriate way and maintain a harmonious balance of water, soil, air, heat, and fertilizer to meet the growth needs of rice. As a special type of soil formed in the process of hydraulic tillage maturation, the fertility of paddy fields is mainly influenced by parent material, climate, hydrology, topography, cultivation practices (especially farming systems), as well as socio-economic, scientific, and technological development. Fertility lays a crucial foundation for paddy field productivity. To achieve high yields in rice production, emphasis must be placed on farmland construction and soil environment improvement, maximizing the potential of paddy soil in fertility contribution to obtain higher yields. The parent materials of soils in paddy fields in South China are diverse and complex, resulting in varying levels of soil fertility. Generally, paddy soils developed from sedimentary materials exhibit better fertility, followed by Quaternary red soil. Under conventional fertilization and management practices, the soil organic matter and total nitrogen of several main farming systems in the double-cropping rice areas of the southern region remain stable and tend to increase. The available phosphorus stably increases, while the available potassium exhibits a clear downward trend. The main reasons behind this are the nutrient removal from crop harvests and fertilization supplements. The fertility balance in paddy fields in the southern region shows an excess of nitrogen and phosphorus and a deficiency of potassium.

The changes in paddy field fertility in the southern region over the past half-century can be divided into four stages based on different cultivation and fertilization practices:

First, in the 1950s and 1960s, conventional rice was cultivated in paddy fields with low multiple cropping indexes and yields. Farmyard manure was primarily adopted for fertilization, which helped maintain the original soil fertility. The soil organic matter content in the first-class fields was above 25 g/kg, total nitrogen was above 1.5 g/kg, and available phosphorus and exchangeable potassium were above 20 mg/kg and 40 mg/kg respectively, ensuring a certain level of reproduction.

Second, in the 1970s, fertilizer-tolerant mid-season hybrid rice and late-season hybrid rice were promoted, leading to an overall increase in fertilizer usage and a higher demand for nitrogen and potassium. With an abundant rural labor force, converting refined and coarse feed into organic fertilizer can be realized by planting green manure in winter, preparing mixed manure in autumn, collecting farmyard manure on a regular basis, and especially focusing on raising pigs. If combined with chemical fertilizers, the soil fertility can also meet the requirements of high-yielding hybrid rice.

Third, in the 1980s, after the reform and opening-up, rural areas implemented the household contract responsibility system, which greatly stimulated the enthusiasm of farmers for farming. China placed great emphasis on grain production, especially in some high-yielding rice regions in the south, the construction of "ton grain fields" was vigorously promoted, and investment was boosted to improve soil fertility. A batch of high-yielding paddy fields was established, such as the first "ton grain city" in Hunan Province—Liling City, with organic matter content in high-yielding fields reaching 51.1 g/kg, total nitrogen 2.9 g/kg, total phosphorus 0.8 g/kg, total potassium 16.6 g/kg, available potassium 118 mg/kg, available phosphorus 7.6 mg/kg and alkaline hydrolysis nitrogen 221.0 mg/kg. However, the soil fertility of paddy fields began to polarize. Some farmers, who had abandoned farming for business activities, engaged in exploitative cultivation practices in paddy fields, prioritizing output over soil fertility.

Fourth, since the 1990s, agricultural high-tech has gained increasing attention, with the new combination of two-line and three-line hybrid rice being highly promoted. In economically developed areas of the southern rice-growing regions, rural land has been transferred to skilled farmers and large-scale growers, which optimizes the allocation of production factors and resources, enhances the production capacity of paddy fields, and further fertilizes the soil. For instance, in Houjing Village, Dongyuan Town, Longhai City, Fujian Province, the average yield over three years for early-season rice was 650.8 kg/*mu*, and for late-season rice, it was 548.7 kg/*mu*. The topsoil contains an organic matter content of 37.4 g/kg, total nitrogen of 1.96 g/kg, available phosphorus of 13 mg/kg, and available potassium of 73 mg/kg. However, with many rural laborers in the southern region and even across the country migrating to towns and other industries, some paddy fields have experienced abandonment because of improper management, resulting in a decline in paddy field production capacity.

To sum up, the increased input of nitrogen fertilizer in rice production in the southern region exceeds the amount of nitrogen taken away by its output, showing a significant increase in the total nitrogen content in rice soil. However, 14.4% of the soil still has a medium-low level of total nitrogen content. In nitrogen fertilizer application and management in rice production, it is necessary to provide classification guidance based on the soil nitrogen content. Strict control of nitrogen fertilizer usage is especially needed for paddy soil with high nitrogen levels to avoid resource waste and exacerbation of non-point source pollution caused by excessive nitrogen fertilizer. In the past two decades, a significant amount of phosphorus fertilizer has been applied in paddy fields in the southern region; thus, the accumulation of phosphorus in paddy soil is quite considerable. Apart from increasing phosphorus fertilizer input in soils with medium-low phosphorus content, phosphorus fertilizer input should be controlled in other paddy fields in phosphorus fertilizer management. In the past two decades, the available potassium content in paddy soils has remained relatively stable, while the total potassium content has shown a trend to decrease, indicating that as the productive level improves, the potassium taken away from the farmland due to rice grain production has not been adequately compensated. Rice cultivation, especially the high-yielding cultivation of super hybrid rice, requires special attention to potassium fertilizer. It is worth noting that farmers in the southern region have a long-standing habit of overapplying nitrogen fertilizer. Phosphorus is easily fixed in the soil, potassium is prone to leaching, while hybrid rice requires a large amount of potassium uptake. Therefore, a nutrient imbalance and a decline in soil fertility occur, which should be given much attention in cultivating high-yielding super hybrid rice.

II. Characteristics of High-Yielding Paddy Field Fertility

The growth period of mid-season super hybrid rice lasts 135 to 155 days, and 45 to 55 days are needed for the rice to transfer from the booting stage to maturity. It is generally asked to "grow well in the early stage, maintain stability in the middle stage, and ensure firmness in the later stage." The yield of rice depends on the soil largely, with soil fertility contributing more than 60% of the yield. Particularly in the middle and later stages, it relies heavily on the soil's capacity to provide continuous fertilizer. Only when deep soil layers serve as a nutrient reservoir can the roots be deeply rooted, exhibiting vitality, ensuring resilience and sustained growth, and maintaining coordination between aboveground and underground growth. This enables them to support super-high yields and withstand various adversities (such as high temperatures, drought, sudden temperature changes, strong winds, and heavy rain), known as "deep roots ensure lush foliage, a solid foundation brings flourishing branches."

The soil profile characteristics of super-high-yielding fields are as follows: the soil exhibits a fine structure with distinct soil profile layers (A–P–W–C), and the absence of adverse structural layers indicates the coordination of water, nutrients, air, and heat. With a depth of 20 to 25 cm (or up to 30 cm for sandy loam soil), the tillage layer (A) is loose and fertile, with mild texture ideally consisting of 30% sand and 70% clay, resulting in good structure, high permeability, and rich plant nutrients. The plow bottom (P) is well-developed and compact, with an optimal thickness of about 8 cm. It has good water retention and nutrient retention capabilities, as well as certain water permeability. The mottled layer (also the accumulation layer) (W) is approximately 50 cm thick, with visible joints and minimal seepage. The subsoil layer (C) is compact. Adverse layers, such as the gray layer, albic layer, and gravel layer, are located 100 cm below the surface.

Nutrient levels and environmental param of super-high-yielding paddy fields are determined as follows. Soil nutrients are sufficient and well-balanced. The soil in high-yielding paddy fields is slightly acidic or neutral, with organic matter content ranging from 25 to 45 g/kg, total nitrogen from 1.5 to 4.1 g/kg, total phosphorus (P_2O_5) from 0.5 to 1.5 g/kg, and total potassium (K_2O) from 15 to 30 g/kg. The available nitrogen is at least 100 mg/kg, available phosphorus 15mg/kg, and available potassium 100 mg/kg. Additionally, the soil has a higher cation exchange capacity (not less than 20 mg equivalents per 100 g of soil) and a higher base saturation degree (60% to 80%), with sufficient primary nutrients and no deficiency in trace elements. There are both active and non-active available nutrients, ensuring a continuous supply of nutrients throughout the growth period without deficiency or excessive absorption, which allows robust and healthy growth of rice plants. Soil redox potential is an indicator that reflects the oxidation-reduction status in soil solution, expressed by Eh in millivolts (mV). The level of soil redox potential depends on the relative concentrations of oxidized and reduced substances in the soil solution and is influenced by factors such as soil aeration, moisture conditions, root metabolism, and the content of easily decomposable organic matter. It reflects changes in the physical, chemical, and biological characteristics of the soil. In a paddy field, the normal Eh during dry periods ranges from 200 mV to 750 mV. When it exceeds 750 mV, the soil is in a completely oxidized state, resulting in rapid organic matter depletion and loss of effectiveness of certain nutrients. In such cases, appropriate irrigation is needed to reduce the Eh levels. During the flooding period, soil Eh in the paddy field fluctuates greatly and can drop as low as −150 mV or even lower. During the drainage and drying period, soil aeration improves, and the Eh increases to over 500 mV. In general, the suitable range of Eh for paddy fields is 200 to 400 mV. When the Eh frequently stays below 180 mV or drops below 100 mV, it indicates that the paddy field is in a strongly reduced state with excessive soil moisture and poor aeration, which hinders the tillering or development of rice. If the Eh remains below −100 mV for a long time, rice will be severely affected by toxic reduction substances and may even die. In such cases, drainage and soil drying should be conducted to increase the Eh value.

To enrich and balance soil nutrients, continuous soil fertilization is necessary, and the key lies in applying a large amount of organic fertilizer to improve the physical and biological characteristics of the soil. When applying chemical fertilizers, attention should be paid to supplementing the deficient nutrients in the soil. The soil texture should be loamy, with a moderate silt-to-sand ratio that is neither too compact nor too sticky, ensuring good tillage properties. The soil porosity should be between 63% and 67%, with a ventilation porosity of 12% to 15%. During the flooding period, the groundwater level should be below 70 cm. There should be a vigorous activity of beneficial microorganisms in the soil and there should be no significant hindering factors such as mineral toxicity, cold waterlogging, hydrogen sulfide, sub-iron hazards, or deficiencies in essential elements (such as zinc and magnesium). Many microbial communities capable of adapting to paddy field environments still thrive in paddy field soils. Microorganisms play an important role in creating and regulating soil fertility. Although the microbial communities in different types of rice soils are not the same, and their mechanisms of action vary significantly, high-yielding paddy fields harbor a multitude of beneficial microorganisms. For instance, the reddish patches on the structural surfaces of paddy soils directly reflect the vigorous activity of beneficial microorganisms in the soil.

To achieve super-high yields in cultivating super hybrid rice, it is advisable to select paddy fields with high-yielding potential. Taking Hunan Province as an example, the most suitable fields for cultivating super hybrid rice are fertile loamy fields in mountainous valleys, hilly basin areas, river basins, and lakeside plains. For paddy fields that do not meet the requirements for super-high yield, the following measures for high-standard farmland construction should be adopted: It is important to establish support systems for ditches, canals, fields, forests, and roads to prevent soil erosion, pollution, ensure smooth irrigation and drainage, level land, and guarantee successful harvests in both drought and flood conditions. In autumn (or winter), plow the fields and expose them to the sun to dry. Deepen the tillage layer each year and mix foreign soil with sand to improve the soil texture. Carry out water-upland rotation and plant leguminous green manure. Effectively return crop straw to the field. Develop animal husbandry and feed production industry and convert livestock and poultry waste to land fertilization. Apply bio-organic fertilizers to improve the physicochemical properties of soil, thus increasing its yield potential to around 900 kg/*mu*.

III. Classification and Function of Fertilizer

As the saying goes, "Crops are the food of human beings, and fertilizer is the food of crops." Apart from meeting the physiological needs and supporting the development of super hybrid rice, fertilizers can also replenish soil nutrient depletion and alter the growth environment, significantly influencing yield and quality. There are various classification methods for fertilizers, and the focus will be on introducing the practice of cultivating super hybrid rice.

1. Main categories and characteristics of fertilizers

(1) *Organic fertilizer.* Organic fertilizer boasts a large amount of organic matter, which needs to be decomposed by microorganisms before being absorbed by plants, resulting in slow and long-lasting fertilization effects. It contains a wide range of essential nutrients but at a relatively low concentration. It contains certain bioactive substances, exhibits obvious effects in soil conditioning and improvement, and is effective in activating or fixing soil nutrients. Examples of organic fertilizers include compost, stable manure, cake fertilizer, human and animal manure and green manure, and they are also known as farmyard manure or slow-release fertilizers.

(2) *Microorganism fertilizer.* Microorganism fertilizer is a biological fertilizer made from one or several beneficial microorganisms, culture substrates, and additives (carriers). It contains highly efficient and active bacteria, and the fertilizer's effectiveness relies on creating suitable conditions

for the growth and reproduction of the microbial strains. The application amount is relatively small, and the method and time of application should be determined by the types and living habits of the microbial strains. Microorganism fertilizer is also known as bio-fertilizer or bacterial fertilizer. Some are referred to as microbial activity fertilizers. Microbial fertilizers do not have direct fertilizing effects on their own (excluding additives), they act as indirect fertilizers by utilizing specialized bacteria's physiological metabolic activities to decompose mineral nutrients in the soil, improve fertilizer efficiency, and fix nitrogen in the air or accelerate the decomposition of organic matter into fertilizers that can be absorbed and utilized by crops, such as silicate bacterium fertilizer (also bio-potassium fertilizer), bio-phosphate fertilizer, biological azotobacter fertilizer, Fuganling, and fermented bacterial fertilizer. Microorganism fertilizers contribute to environmental protection, resource conservation, as well as the benign cycle of ecology.

(3) *Fertilizer.* Chemical fertilizer, abbreviated as fertilizer, is a type of fertilizer that contains one or several essential nutrients for crop growth, which are produced by chemical and/or physical methods. Chemical fertilizers have relatively simple compositions and higher nutrient concentrations. They are mostly water-soluble or acid-soluble, belonging to fast-acting nutritional substances that can be directly absorbed by the roots or foliage. After being applied to the soil, they can regulate the concentration of the nutrient elements and the physicochemical properties of soil, such as ammonium bicarbonate, urea, potassium chloride, and superphosphate. They do not contain organic matter and are also known as mineral fertilizers, inorganic fertilizers, commercial fertilizers, or quick-release fertilizers. Based on the composition formula, they can be further divided into single-element fertilizers and compound fertilizers (including compound blended fertilizers, compound formulated fertilizers, multi-functional fertilizers, and slow-release fertilizers). Chemical fertilizers can be categorized based on the amount absorbed by plants into three groups: macro-element fertilizers (e.g., nitrogen, phosphorus, and potassium), which are essential in large quantities, particularly for crops like rice that require substantial amounts of silicon; medium-element fertilizers (e.g., sulfur, calcium, and magnesium), which are needed in moderate quantities; and trace element fertilizers (e.g., iron, manganese, boron, zinc, molybdenum, copper, chlorine, and rare earth fertilizers). The common characteristics of chemical fertilizers are high nutrient concentrations and quick efficacy. They will have an immediate effect once applied. However, improper use can lead to side effects, especially when it comes to trace element fertilizers. Even a slight excess can bring about severe toxicity. Therefore, the dosage and application should be carefully considered, and the scientific approach is to apply fertilizers based on soil testing or physiological determination indicators.

(4) *Green manure.* Green manure is an organic fertilizer derived from green plants. It is a nutrient-rich biological fertilizer and ranks as one of the important traditional organic fertilizers in China. Due to its unique role, it is classified separately from other organic fertilizers. Green manure has the following characteristics:

a) Green manure comes in various types and has strong adaptability, making it easy to cultivate. In terms of sources, it can be classified into cultivated green manure and wild green manure. In terms of botany, it can be classified into leguminous green manure and non-leguminous green manure. In terms of the planting seasons, it can be classified into winter green manure, summer green manure, and perennial green manure. In terms of utilization methods, it can be classified into field green manure, cover green manure, green manure for vegetable and crop rotation, green manure for livestock feed, and green manure for grain production. In terms of growing environment, it can be classified into dryland green manure and aquatic green manure.

b) By combining cultivation and land management, green manure can improve soil quality and fertility. Green manure, with its abundant foliage covering the ground, can prevent or reduce the loss of water, soil, and nutrients. It contains a large amount of organic matter, which undergoes continuous decomposition under the action of microorganisms when incorporated into the soil. It not only releases available nutrients but also forms humus. The combination of humus and calcium creates a granular soil structure, which is loose, well-drained, and has strong water and nutrient retention capabilities. Such soil with a granular structure exhibits excellent performance in regulating water, nutrients, air, and temperature. Conversely, green manure can promote rapid reproduction and activity of soil microorganisms, facilitating the formation of humus, accelerating soil maturity, and enhancing nutrient availability. Green manure crops contain nutrients such as nitrogen, phosphorus, potassium, and various trace elements. The secretion of green manure crops during their growth and the organic acid of decomposition after plowing can transform insoluble phosphorus and potassium into available forms that can be effectively utilized by crops, thus ensuring stable fertilizer efficacy. Every 1,000 kg of fresh green manure grass provides approximately 6.3 kg of nitrogen, 1.3 kg of phosphorus, and 5 kg of potassium, which is equivalent to 13.7 kg of urea, 6 kg of calcium superphosphate, and 10 kg of potassium sulfate. The yield of fresh green manure grass is between 1,000 to 3,000 kg/*mu*, which reduces the use of chemical fertilizers and promotes the development of green agriculture and organic agriculture.

c) Little investment and low cost. Green manure requires only a small number of seeds and fertilizers. It can be planted on-site and applied on-site, saving labor and transportation costs compared with chemical fertilizers.

d) Comprehensive utilization results in great benefits. Green manure can be used as feed for livestock, promoting the development of animal husbandry. In turn, the livestock manure can fertilize the fields, resulting in a mutually beneficial relationship. Green manure can also be used as raw material for biogas, contributing to the solution of certain energy issues. Biogas fertilizer is an excellent organic fertilizer. Green manure, like Chinese milk vetch, serves as a good source of nectar and can be utilized for beekeeping.

Super rice fields are used to plant leguminous green manure Chinese milk vetch, which is also called genge or lacy phacelia. The fresh grass of Chinese milk vetch contains 0.4% nitrogen (N), 0.11% phosphorus (P_2O_5), and 0.35% potassium (K_2O), while the dry grass of Chinese milk vetch contains 24% crude protein, 4.7% crude fat, 15.6% crude fiber, and 7.6% of ash content. It is not only a high-quality organic fertilizer but a good feed for livestock. Chinese milk vetch boasts a strong nitrogen-fixing ability, and its rhizobia is sensitive to phosphorus. It is recommended to apply phosphorus and nitrogen during the early stage of growth to enhance phosphorus and nitrogen availability, allowing for greater nitrogen yield and increased overall biomass production. Another type of green manure is field radish, also known as garden flowers. It is a cruciferous plant with an erect stem. Typically, Chinese milk vetch is mixed with radish seeds and/or rape seeds and then sown successively (to adjust the sowing period), which allows each crop to leverage its advantages during the growth process and after incorporating it into the soil, thus increasing production and improving soil fertility.

2. Main fertilization methods and characteristics

(1) *Base Fertilizer.* Base fertilizer, also known as bottom fertilizer, refers to the fertilizers applied to level the land before rice sowing or transplanting. It serves to establish a nutritional foundation for seedling emergence, root development, tillering, and the whole growth period. It allows nutrients to be stored in the soil, achieving the effects of "fertilizer fertilizes soil, soil nourishes seedlings, and strong seedlings produces strong grains." Base fertilizer can be achieved through whole-layer application or surface application, and the types of fertilizers include organic fertilizers (including green manure), inorganic fertilizers and bio-fertilizers, with organic fertilizers being the primary choice.

(2) *Topdressing.* It refers to the application of fertilizers after rice sowing or transplanting to supplement the nutritional needs of different growth stages, regulate the growth and appearance of the seedlings, and ensure their vigorous growth until achieving high-quality and high-yielding production. The types of top-dressing fertilizers include inorganic fertilizers and organic fertilizers, with inorganic fertilizers (quick-release fertilizers) being the primary choice. Depending on the specific growth stage and desired effect, top-dressing fertilizers can be further divided into different categories with specific names. For example, during the seedling stage, there are weaning fertilizers and initial growth fertilizers; during the tillering stage, there are tillering fertilizers, elongation fertilizers, panicle fertilizers (which can be further divided into flowering-promoting fertilizers, and flower-preserving fertilizers), and grain-filling fertilizers. The fertilizer is also referred to as balanced fertilizer when the purpose of topdressing is to promote balanced growth. According to the targeted areas for topdressing, it can be further categorized as deep-layer application (such as deep placement of basal fertilizer or ball-shaped fertilizer), surface application (such as applying near the root zone), and external root topdressing (such as foliar fertilizer).

In addition, seed fertilizers (such as humic acid fertilizer mixed with seeds) and seedling root fertilizers (using phosphorus fertilizers or rare earth elements mixed into a slurry to coat the roots) can also be classified into base fertilization and topdressing. These methods are convenient and can be adopted proactively. The purpose is to use small amounts of fertilizer to support the growth of larger amounts of fertilizer, reducing labor and cost. The specific application depends on the circumstances.

IV. Nutrient Absorption and Fertilization Requirements of High-Yielding Rice

The nitrogen uptake per unit yield of rice increases correspondingly with the increase in yield. Studies have found that when the yield reaches 500 kg/*mu*, the nitrogen uptake per 100 kg of rice grain produced is about 1.85 kg. When the yield rises to over 800 kg, the nitrogen uptake per 100 kg of rice grain increases to around 2.18 kg. This is because the nitrogen content in rice plants from booting to maturity increases with the increase in yield. The higher nitrogen content in the later stages of the plant is beneficial to the improvement of photosynthetic productivity, which is a characteristic of nitrogen uptake in high-yielding rice. The nitrogen absorption of rice populations at different yield levels during different growth stages reveals that the increase in nitrogen uptake does not occur in the tillering or elongation stage but primarily from the elongation stage to the booting stage and, to a lesser extent, from the booting stage to maturing stage. Therefore, efforts to achieve high yields should focus on improving the nitrogen uptake capacity of the population from the elongation stage to the booting stage, as well as from the booting stage to the maturing stage. All this asks us to "control at first, and then increase the yield" in fertilizer management, focusing on the application of nitrogen fertilizer after the elongation stage. Meanwhile, the phosphorus uptake of rice increases with the increase in yield. High-yielding rice exhibits a significant increase in phosphorus uptake from the elongation stage to the booting stage. Even during the booting stage to the maturing stage, high-yielding rice still displays relatively high phosphorus uptake. To achieve high yields, it is important to increase the application of phosphorus fertilizer in the middle and later stages. The potassium uptake of rice increases greatly with the increase in yield level, showing a positive correlation between yield and potassium uptake. To cultivate high-yielding rice, it is important to increase the application of potassium fertilizer, particularly in the middle stage, which helps achieve a higher potassium content during the booting stage and meet the physiological and metabolic needs of super hybrid rice.

The fertilizer application amount should be determined by the total nutrient requirement for producing 800 to 900 kg of rice per *mu*, as well as the specific fertilizer needs of super hybrid rice and different ecological soil regions, and then carry-on scientific formulation and adopt effective methods. Specific operational techniques are as follows:

In the middle and lower reaches of the Yangtze River, single-cropping super hybrid rice targets a yield of 800 to 900 kg of rice per *mu*. The yield is determined by the field fertility, and the fertilizer application is determined by the yield. According to the absorption ratio of nitrogen (N), phosphorus (P_2O_5), and potassium (K_2O) by high-yielding rice, which is 1:0.5:1.1, and based on the average nitrogen requirement of 2.0 kg per 100 kg of rice, the fertilizer application amounts needed are approximately 16 to 18 kg/*mu* for nitrogen, 7.2 to 8.1 kg for phosphorus, and 19.2 to 21.6 kg for potassium. This is an empirical formula based on scientific experiments and long-term cultivation of such paddy fields, which considers the offset between soil nutrient-supplying capacity (soil fertility contribution) and the fertilizer utilization rate during the current season. Specific fertilizer application rates may need to be adjusted according to the variations in soil fertility under different ecological conditions. If the soil productivity contribution exceeds the unutilized fertilizer rate, the fertilizer application can be reduced; otherwise, it should be increased. It is particularly emphasized that high-yielding cultivation should prioritize the application of well-rotted organic fertilizer, supplemented by chemical fertilizers. The ratio of base fertilizer to topdressing generally varies, such as 5:5, 4:6, or 6:4, depending on growth characteristics, soil fertility, and fertilizer properties. Topdressing can be further divided into tillering fertilizer and panicle fertilizer, with ratios of 5.5:4.5, 6:4, or 7:3, which are determined by the variety, soil fertility, and ecological characteristics. With nitrogen fertilizer as the main component, long-lasting organic fertilizer, and slow-release phosphate fertilizer as base fertilizer, potassium fertilizer is combined with nitrogen fertilizer (including compound fertilizer) to be applied in multiple stages. The main fertilizer application measures to achieve a yield of 800 to 900 kg of rice per *mu* are as follows:

1) *Apply sufficient base fertilizer.* Apply 50–80 kg of vegetable residues per *mu*, 25–30 kg of calcium-magnesium phosphate fertilizer, 11.7–15 kg of potassium chloride, 40 kg of compound fertilizer (with an N:P:K ratio of 18:10:17), and 1.5 kg of zinc fertilizer. The base fertilizer should be applied three to five days before transplanting. Mix the various fertilizers with vegetable residues and spread them evenly in the plowed and harrowed field. Then, use agricultural machinery or animal-powered harrows to level the field, maintaining a shallow water level to preserve fertility before planting.
2) *Apply early tillering fertilizer.* The rice seedlings will stabilize and begin tillering seven to eight days after transplanting. It is necessary to promptly apply tillering fertilizer, with a recommended application rate of 7–9 kg of urea per *mu*.
3) *Apply panicle fertilizer appropriately.* During the second irrigation after the field has been dried, rice plants enter the stage of stem elongation, a critical period for panicle differentiation. When the young panicles are at the stage of differentiation II–III (around the time of the third-to-last leaf showing a tip until the middle stage), timely application of panicle fertilizer is recommended to promote differentiation of branches and spikelet. It is advised to apply 4 to 6 kg of urea and 7.5 to 10 kg of potassium chloride per *mu*. In the early stages of panicle differentiation IV–V (around the time of the second-to-last leaf showing a tip until the appearance of the first awn), if there are signs of yellowing leaves in the field or insufficient nutrient supply, additional flower-preserving fertilizer should be applied.
4) *Apply grain-strengthening foliar fertilizer.* To prolong the lifespan of functional leaves and ensure grain weight, spray 50 g of grain extract diluted in 50 kg of water per *mu* onto the leaves during the initial heading stage and heading stage. This increases the grain-setting rate and enhances grain weight. It also boasts a good protective effect against adverse weather conditions, such as low-temperature damage.

In tropical ecological regions (such as Sanya in Hainan Province), it has been found from collected data and recent practices that super hybrid rice can achieve high yields even in areas with medium or low soil fertility (such as sandy soil with poor nutrient retention, where nutrients are easily lost during the rainy season and decompose quickly during the dry season). In such conditions, for super-high-yielding cultivation, it is recommended to apply 30 kg of pure nitrogen, 15 kg of phosphorus (P_2O_5), and 30 kg of potassium (K_2O) per *mu* every three months.

Although the cultivation seasons for single-cropping super hybrid rice in Hunan and Hainan are different, successful yields of 800 to 900 kg/*mu* can be achieved. Examples of fertilizer management measures indicate that they are fundamentally like the four points mentioned above: First, sufficient application of base fertilizer on the entire soil layer. Apply 1,000 kg of pig and cow manure, 30 kg of phosphorus fertilizer, and 40 kg of compound fertilizer (45% NPK) per *mu* as base fertilizer. After the first plowing (three to five days before transplanting), mix and spread the fertilizer, then perform harrowing to ensure deep fertilization throughout the soil layer, which effectively prevents phosphorus fixation and nitrogen-potassium loss and improves fertilizer utilization. Second, early and frequent application of tillering fertilizer. After transplanting, the rice seedlings will stabilize and start tillering within five to seven days. After the appearance of tillers, it is recommended to apply 10 kg of urea and 10 kg of potassium fertilizer per *mu* every three to four days. Considering the poor nutrient retention of sandy soil, the tillering fertilizer should be applied several times in appropriate amounts, combined with irrigation and fertilization practices alternating between wet and dry conditions. Third, timely and moderate application of panicle fertilizer. During the booting stage II–III, timely application of panicle fertilizer should be carried out to promote the growth of grains, and the specific application rate should be determined by the growth situation of the seedlings. Fourth, foliar application of grain-strengthening fertilizer. To prolong the lifespan of functional leaves and ensure grain weight, spray a mixture of 100 g of potassium dihydrogen phosphate, 50 g of urea, and 50 kg of water onto the leaves during the heading stage and milk ripening stages. Applying the foliar fertilizer before 10:00 a.m. or after 3:00 p.m. is advisable.

V. The Fertilization Amount and Approaches for Rice Breeding

The nutrients absorbed by rice come from two sources: the soil itself and the fertilizers applied in season.

1. Test soil for formulated fertilization

Experiments confirm that there is a certain correlation between soil nutrient determination values and the nutrients absorbed by crops. Based on observations in production practices and fertilizer comparison experiments, it is estimated that the soil itself can provide about 1/2 to 2/3 of the nutrients required for rice production. However, the supply situation varies greatly from region to production level. When assessing the fertility status of a specific region's soil, it is essential to conduct soil analysis and experiments based on the relationship between local soil properties, fertilization levels, and yields, which helps to determine soil fertility more accurately. Choose testing methods with high relevance for nutrient testing, while setting up field biological experiments in the plot with complete fertilization zones (NPK zone) and deficient zones (NP zone, NK zone, PK zone, and a zero-fertilization zone). Calculate the yield after harvest and express the abundance or deficiency of soil nutrients based on the percentage of yield in the deficient nutrient zones compared to the complete fertilization zone. Classify nutrient test values based on crop yield levels. According to general standards in China, categorize soil nutrient test values below 50% of the relative yield in the complete fertilization zone as extremely low, 50%–75% as low, 75%–85% as moderate, 85%–95% as high, and above 95% as extremely high. This helps to determine the soil nutrient deficiency indicators suitable for a particular rice variety, create a nutrient deficiency index, and establish a table for fertilizer application based on nutrient levels. When a certain soil nutrient determination value is obtained, the retrieval table allows users to understand the abundance or deficiency of soil nutrients and the approximate range of fertilizer application quantities. This serves as the basis for formula fertilization, while also considering other influencing factors on rice field productivity and yield goals to determine the maximum and optimum fertilizer application rates.

2. Target production

The target production should be decided by soil fertility because it is the foundation of the yield. Typically, the empirical formula $y = a + bx$ (simple linear regression equation) represents the relationship, where “a” refers to the production in the zero-fertilization zone (base yield), and “b” refers to the slope. Under different soil fertility conditions, the production varies according to the changes in fertility. The varying soil fertility is considered as the independent variable (x). Starting from the comparison between yields in the zero-fertilization zone and their corresponding soil fertility, the highest attainable yield (y) is determined by field experiments and identified as the target yield. Alternatively, the average yield of the previous three years’ crops in the local area can be used as a basis, and an additional 10% to 15% increase in yield can be set as the target production.

3. Fertilizing utilization rate

The fertilizer utilization rate has a significant impact on the accuracy of fertilizer quantification during the growing season. However, the fertilizer utilization rate is not constant and is influenced by various factors such as crop variety, soil fertility, fertilizer quantity, and fertilization methods. Therefore, within each fertilization zone, field experiments should be conducted to determine the fertilizer utilization rate for the local area with the difference reduction method formula. Alternatively, existing research data on fertilizer utilization rate can also be considered. In the major rice-growing regions in the southern part of China, the nitrogen (N), phosphorus (P), and potassium (K) utilization rates are approximately 30%, 20%, and 40%, respectively. Higher fertilizer utilization rate in the growing season results in higher yields, with nitrogen fertilizer utilization rate in super-high-yielding paddy fields reaching up to 51%.

4. Calculation of fertilizer application amount

The fertilizer absorption required for the yield can be estimated as 1 kg of pure nitrogen is equivalent to producing 50 kg of rice. Considering the fertilizer supplied by the soil and the fertilizer utilization rate for the current season, the calculation can be estimated with the following example.

For example, if a paddy field plans to produce 550 kg of rice, the natural fertility yield (namely, yield without fertilizer) is 350 kg (contribution rate of soil fertility is 63.3%). Deducting the natural fertility yield, 200 kg of yield will be achieved with fertilizer application, which requires 4 kg of pure nitrogen (converted based on the aforementioned indicator). Considering the fertilizer utilization rate of nitrogen is 35%, then approximately 11.4 kg of pure nitrogen will be needed (the unutilized portion of fertilizer is 65%, which is equivalent to the contribution rate of soil fertility). Then, based on the nitrogen content of various nitrogen fertilizers, the application quantities of each nitrogen fertilizer can be determined. The application amount of phosphorus fertilizer (P_2O_5) is approximately half of the pure nitrogen for early-season rice and around 40% of the pure nitrogen for late-season rice. As for potassium fertilizer (K_2O), the quantity applied is equivalent to that of nitrogen.

5. Balanced fertilization and formulated fertilization

Balanced fertilization refers to the balanced supply of various nutrients required by crops from both the soil and applied fertilizers, meeting the needs for crop growth. Formulated fertilization is a method of precise scientific fertilization based on the understanding of crop nutrient requirements, soil fertility conditions, and fertilizer efficiency. In practice, regardless of whether it is balanced fertilization or formulated fertilization, the goal is to overcome blinded application and low efficiency, while improving the scientific validity, predictivity, and efficiency. The basic principle remains the same: optimizing the combination of fertilizers based on the crop's target yield, soil fertility levels, and the various available fertilizer options in accordance with the crop's specific nutrient requirements.

VI. Formula Fertilization Technology for Super Hybrid Rice

Based on the fertilization requirements of super hybrid rice and the soil characteristics of different ecosystems, the total nutrient quantity and fertilizer application amount needed to achieve a yield of 900 kg of rice per *mu* for super hybrid rice is determined. The following is an example of the successful implementation of specific fertilization technology by the Hunan Hybrid Rice Research Center in Longhui County, Hunan Province, for achieving super-high yields (900 kg/*mu*) of super rice.

Application of single-cropping super rice with a yield of over 900 kg/*mu* in Longhui County, Hunan Province indicated key features as follows: For high-yielding rice, the nutrient absorption of nitrogen, phosphorus, and potassium should be divided into stages during its growth and development. Multiple applications of scientifically balanced formulated fertilizers should be used. In addition to nitrogen fertilizer, phosphorus and potassium fertilizers should be applied reasonably. Based on soil fertility and fertilizer utilization rate, approximately 23 kg of pure nitrogen per *mu* are required for a yield of 900 kg, with an $N:P_2O_5:K_2O$ ratio of 1:0.6:1.1. Sufficient base fertilizer, early tillering fertilizer, timely and appropriate application of panicle fertilizer, and late-stage application of grain fertilizer should be applied.

1) *Apply sufficient base fertilizer.* Base fertilizer, also known as bottom fertilizer, provides the basic nutrients for the whole growth period of rice and must be applied adequately to establish a solid foundation for high yields. The base fertilizer accounts for about 40% of the total fertilization amount and is generally applied once before transplanting or plowing. For each *mu*, 50 kg of compound fertilizer with a nutrient content of 45% (15-15-15) and 200–500 kg of organic fertilizer is adopted.
2) *Apply early tillering fertilizer.* Early tillering fertilizer accounts for about 30% of the total fertilization amount. The effective tillering period occurs around 30 days after transplanting. It is recommended to apply quick-release nitrogen fertilizer for early tillering. The first application should be done around 6 days after transplanting, with a rate of 8–10 kg of urea per *mu* and 7.5 kg of compound fertilizer with a ratio of 45% (15-15-15). The second application should be made around 14 days after transplanting, with a rate of 3–5 kg of urea per *mu* and 7.5 kg of potassium fertilizer, which promotes early, quick, and prolific tillering in the seedlings.
3) *Apply panicle fertilizer in a timely and appropriate method.* Panicle fertilizer accounts for about 30% of the total fertilization amount and serves as the foundation for promoting large panicles and high yields. Apply panicle fertilizer when investigating the differentiation of the main-stem panicles at the second stage. Adopt a rate of 4–6 kg of urea per *mu* and 7.5–10 kg of compound fertilizer with a ratio of 45% (15-15-15). Additionally, apply 8–10 kg of chloride fertilizer as a flower-promoting fertilizer. Around the fourth stage of main-stem panicle differentiation, apply 3–5 kg of urea per *mu* and 7.5 kg of chloride fertilizer as a flower-preserving fertilizer. The application of panicle fertilizer serves two purposes: provide sufficient nutrients during the entire panicle-differentiation stage to promote the differentiation of branches and spikelets, while preventing degradation due to nutrient deficiency and laying the foundation for the formation of large panicles; provide nutrients to the top three functional leaves to extend their functional period, which plays a role in root development, leaf protection, stem strengthening and lodging prevention.
4) *Apply grain fertilizer as a supplementary application in the later stages, with foliar fertilizer being the main type.* Super rice boasts large and abundant grains. If the management of fertilizer and water is not appropriate, it is prone to cause nutrient deficiency and premature senescence in the later stages. Therefore, disease and pest control should be conducted in the later heading stage. Spray 50 g of urea and 1 packet of grain booster (50 g/packet) diluted in 60 kg of water per *mu* on the leaves to promote uniform panicle development, reduce empty panicle rate, and improve grain weight.

CHAPTER 5

Water Management Technology to Achieve Ultra-high Yield of Super Hybrid Rice

I. Water Requirement Rules of Super Hybrid Rice

The water requirements of rice can be divided into physiological requirements and ecological requirements. Physiological water requirement refers to the water needed for the growth, development, and normal life activities of rice plants, including the evaporated water of rice plants and the water that makes up the rice plant bodies. Ecological water requirement refers to the water required to create a favorable ecological environment to ensure the growth and development of rice, including the water of inter-row evaporation and water leakage from paddy fields.

In the southern rice-growing areas, the water requirement for mid-season hybrid rice seedbeds is 85–180 mm, and for primary farmland is 540–770 mm. The water requirement for single-cropping late-season rice seedbeds is generally the same as mid-season hybrid rice, and the water needed for primary farmland is 330–690 mm. During the growth and development process of rice, the water requirement increases first and then decreases. For single-cropping mid- and late-season rice varieties, the water requirement accounts for 5.7%–12.9% during the transplanting and regreening stage, 25.1%–26.3% during the tillering stage, 24.3%–35.1% during the elongation-booting stage, 9.4%–17.9% during the heading stage, 9.5%–13.5% during the milk ripening stage, and 6.1%–9.9% during the maturing stage. The single-cropping mid- and late-season hybrid rice plants are taller with larger leaves, and their water requirement is higher than that of double-cropping early- and late-season hybrid rice. In the early stage, when the temperature is relatively high, the water demand rises rapidly. The peak period occurs earlier, mostly during the elongation-booting stage. The daily water requirement during the primary farmland stage is 4.6–6.0 mm per day, while the water requirement during the peak period is 5.5–7.2 mm per day. The critical period for water in rice is the booting stage. If there is a shortage of water during this period, it can easily result in smaller and fewer grains in the panicles or even cause empty husks. Ensuring water supply during the booting stage is beneficial to the formation of large grains and increases yields.

II. Principles of Water Management for Super Hybrid Rice

For super hybrid rice, water requirements vary at different growth stages. To achieve the goal of producing stable and high-yielding (900 kg/*mu*) super hybrid rice, scientific water management methods should be implemented for different growth stages.

1. Frequent irrigation with thin water during the tillering stage

During the tillering period, implementing thin and frequent irrigation is beneficial to increase water and soil temperature, enhance oxygen in the soil, promote healthy root development, improve nutrient absorption capacity, stimulate early tillering, and increase the panicle formation rate. The water management after transplanting varies from the planting methods. For hand-transplanted seedlings, a shallow water layer of 3–4 cm should be maintained. For seedlings thrown into the field, a water layer of 2–3 cm should be maintained. For machine-transplanted seedlings, a shallow water layer of 1.5–2 cm should be irrigated. At the same time, the water layer should also be adjusted to the weather, with a depth generally around 3 cm, shallower in cloudy and rainy weather and deeper in hot and dry weather.

2. Drain and sun-dry the field timely during the seedling stage

In the later stage of tillering, when the total number of seedlings per *mu* of super hybrid rice reaches about 180,000, timely drainage and field-drying should be carried out to control ineffective tillering, which is beneficial for the main stem and large tiller to grow rapidly and achieve the goal of multiple and large panicles. When drying the field, it is important to consider the weather conditions. In rainy weather, it is difficult to sun-dry the fields, so the sun-drying time can be longer. It is important to take advantage of sunny days for early sun-drying or seize any opportunity of sunny intervals for sun-drying. In dry weather or areas with poor water conditions or water shortages, fields can be lightly sun-dried or left without sun-drying.

3. Frequent irrigation with shallow water during the panicle differentiation stage

If there is insufficient water during the stage of panicle differentiation, it will reduce the amount of spikelets, resulting in spikelet degeneration and a decrease in grain number. If there is too much irrigation, it will make the rice stems soft and lodging. Generally, it is advisable to maintain a water layer of six to eight cm. In case of drought, the water layer can be slightly deeper. For fields with frequent cloudy and rainy weather, high groundwater levels, or overuse of fertilizer, the dry-focused watering strategy should be applied. Super hybrid rice is not tolerant to low temperatures during the differentiation stage of young panicles. When encountering strong cold air, deep-water irrigation should be carried out to maintain a certain depth of water layer, stabilize soil temperature, and reduce the impact of low temperatures on spikelet development.

4. Maintain a water layer during the booting and early heading stage

During the booting and early heading stage, rice is most sensitive to water, as a water shortage during this period can easily lead to empty husks and yield reduction. It is advisable to maintain a water depth of eight to ten cm during this stage, as this can regulate temperature and increase air humidity between plants. When encountering the Cold Dew wind or low temperatures, deep-water irrigation should be carried out to keep the water layer warm.

5. Alternate dry and wet conditions during the milk ripening stage

The milk ripening stage is a critical period for super hybrid rice grain-filling. In this period, drought and water shortage can lead to premature leaf senescence and decreased photosynthetic efficiency, therefore affecting grain-filling. If the deep-water layer is maintained, the oxygen in the soil will also decrease, affecting the later root activity, causing early leaf senescence, and resulting in poor grain-filling. Therefore, in the milk ripening stage, it is necessary to keep the soil moist. The method of running water should be adopted to ensure that the water is not stagnant and there is no shortage of water. Alternative dry-wet irrigation is beneficial for grain-filling. After the yellow ripening period, the water demand greatly decreases. It is important to avoid stagnant water in the field during this period to promote maturity and facilitate harvesting.

III. The Function and Methods of Sun-Drying the Field

In rice paddy soil with water layer irrigation, the soil is mainly composed of solid phase and liquid phase, with very little air in the soil. After a period of water layer irrigation during the tillering stage, the redox potential of the soil has significantly decreased and toxic reducing substances have increased. After sun-drying the field, air directly enters the soil, gas exchange occurs, soil oxygen content increases, carbon dioxide content decreases, and redox potential increases. Soil loses water and shrinks, micro aggregation increases and soil permeability improves, which is conducive to soil environment renewal, reducing restorative substances and promoting organic matter mineralization. During the drying process, ammonium nitrogen and valid phosphorus content in the soil decrease and increase again after rehydration. Sun-drying has the effect of "controlling first and promoting later."

For rice plants, mid-stage (from the seedling stage to the beginning of elongation) sun-drying and control of soil moisture and nitrogen supply can renew the soil environment and adjust the carbon-nitrogen ratio within the plants. This promotes leaf senescence while controlling ineffective tillering and growth of the base internodes and transitional leaves and stimulating root growth. Improving the plant and optimizing the group structure during the mid-stage of rice growth is beneficial to solve various contradictions between the overground and underground parts of the plant, between rice plants and the environment, and between the panicle number, strong stems, and large panicles. It is one of the most important means of mid-term regulation in high-yielding groups.

In traditional sun-drying methods, the sun-drying process is delayed and applied to a large area. It is hoped that the purpose of controlling growth and improving plant type can be achieved through a one-time thorough-sun-drying. Generally, rice fields are sun-dried until cracks appear on the field edge, with the middle of the field showing chicken-foot-like cracks that allow people to stand without sinking and the leaves visibly turning yellow, known as "thorough-sun-drying." Thorough-sun-dried fields have many adverse effects on plant growth. First, there are soil cracks that break the roots, causing the shedding of lateral roots and root hairs, and even damaging the root epidermis and endodermis, severely weakening the root vitality. The absorption capacity of roots cannot be restored in the short term after rehydration, which cannot achieve the goal of "controlling first and then promoting." Moreover, during the thorough-sun-drying of the field, many new roots (including lateral roots) extend from the soil layer to the surface. These thick white roots are produced under extremely good ventilation conditions. Once flooded, they stop growing, lose vitality, and quickly rot, rendering them unable to absorb nutrients. During the long-term sun-drying process, soil microbial activity is vigorous. After rehydration, due to the oxygen consumption by many microorganisms, the soil rapidly turns into a reduced state, which is unfavorable for root growth. Therefore, mid-term sun-drying should not be done all at once. First, moderate sun-drying takes the initiative in the time of sun-drying. The seedling stage is suitable for sun-drying, and the soil is in a moist state, which can effectively control the ineffective tillering. The panicle formation rate is high. Second, it is conducive to the formation of large panicles. In the fields sun-dried moderately for multiple times, the number of spikelets on the main stem and the first and second branch stems of tillering is higher than those in fields with one thorough-sun-drying compared with the thorough-sun-dried fields, and there are fewer degenerated spikelets. When sun-drying the field, it is necessary to dig drainage ditches in the field, and the depth should be based on the depth of the plow bottom layer.

IV. Super Hybrid Rice Water Management Techniques

In rice production, most farmers adopt flooding irrigation methods, and they often cut off the water supply in the later growth stage. According to a survey conducted on farmers, 76% of farmers use the flood irrigation method, and 57% of farmers experience early dehydration in the later stage of cultivation. Flooding irrigation not only causes fertilizer loss (especially during the fertilizing stage) but also inhibits tillering and root growth, leading to rice blast and stripe disease. Cutting off water too early in the later stage will seriously affect the grain-setting rate and yield.

In recent years, many new concepts and methods have been proposed at home and abroad, such as furrow irrigation, alternative dry-wet irrigation, intermittent-wetting irrigation, and mulching dry-sowing, which have played a positive role in the transformation from traditional water-rich and high-yielding irrigation to water-saving and high-yielding irrigation and improving water use efficiency. In terms of irrigation methods, it has evolved from uniform irrigation to localized irrigation that regulates plant functions and improves water use efficiency. It emphasizes alternating control of dryness in some root zones and moisture in other root zones to regulate stomatal opening to the most suitable level, in order to improve crop water use efficiency without sacrificing crop photosynthetic production.

1. Water management technology for intensive cultivation

The core of intensive cultivation of super rice rests in water-saving irrigation. The whole growth stage should be moist or relatively moist on the field surface to develop roots to achieve the goal of cultivating strong individual plants, enlarging populations, and producing high yields with lodging resistance.

1) *Seedling stage.* Keep the soil moist to promote seedling emergence. In the one-leaf-one-bud stage, control should be emphasized to promote the formation of a strong root system. If it adopts a dry-bed breeding method, there is no need to irrigate; spraying during drought is enough. If it adopts a plastic-tray seedling-breeding method, just keep the tray moist. There

is no need for a water layer on the surface. For a conventional wet-bed breeding method, water in the trench and a moist surface are enough.

2) *Transplanting stage.* Different management measures are taken based on different seedling cultivation methods. For dry-bed or tray seedling cultivation, transplanting with soil is required, and there is no need for surface water in the field, just keep it moist to ensure timely seedling emergence. For conventional wet-bed seedling cultivation, uprooting and transplanting are performed, and after planting, keep the field moist for two to three days to promote regrowth.
3) *Regreening and tillering stage.* Generally, seedlings begin to grow upright and turn green three to five days after transplanting. The field should maintain shallow water, which is beneficial for the rapid and vigorous growth of seedlings. After tillering, the field should be managed with alternating dry and wet conditions, with the soil slightly cracked when dry, and fertilizers and pesticides applied when the water level is shallow to medium. When the number of seedlings in the field reaches 70% to 90% of the target number of valid panicles (can be lower for strong tillering capability and higher for weak capability), it is necessary to sun-dry the field, which controls seedlings, promotes field ventilation, lowers the field humidity, and reduces pests and diseases.
4) *Panicle differentiation stage.* After sun-drying the field and controlling the seedlings, it is necessary to replenish water in a timely manner during the differentiation of young panicles. Intermittently irrigate a shallow water layer of about five cm, ensuring that the water from the last watering is not visible when watering again, with an interval of two to three days, to meet the water requirement for young panicle differentiation.
5) *Heading and grain-filling stage.* Keep the water layer at about six cm to ensure rapid heading. Subsequently, maintain a balance between dryness and wetness, with occasional exposure and irrigation, ensuring no dehydration in three to five days before maturity. Due to the large size of grains, the filling period of super hybrid rice is longer than that of ordinary hybrid rice, so special attention should be paid to its water management before maturity.

2. Precise water management technology for quantitative cultivation

Water management is mainly carried out during the tillering stage and heading and grain-filling stage. For the tillering stage of rice, the soil layer needs to be irrigated frequently after transplanting the mid- and large-size seedlings to ensure that the soil surface has a shallow water layer. For the transplantation of small seedlings, after reaching the one-leaf stage, intermittent watering and exposure should be adopted to promote the development of rice roots. Once the rice reaches the two-leaf stage, irrigation can be carried out by alternating between shallow water layer irrigation and surface exposure. To control the ineffective tillering, the time for sun-drying the field should be set in the two-leaf stage before the ineffective tillering. During the heading and grain-filling stage of rice, the soil should be in a dry-wet-alternating state. When watering, the water layer should be two to three cm deep. After the water layer disappears for about four days, water should be applied again, and the water layer should still be two to three cm deep. In the precise water management of rice, maintaining a water layer of two to three cm after transplanting can effectively reduce the damage caused by the underdeveloped root system and adversity. When the plants gradually grow, and the root system develops, drainage and sun-drying can effectively increase the oxygen content in the soil, promote the respiration of rice roots, reduce the accumulation of toxic substances, and strengthen the support function of the root system that can effectively prevent lodging.

3. Water management technology for water-saving high-yielding cultivation

Implement wet management of air irrigation (mild alternative dry-wet irrigation) effectively to promote the growth of rice roots. During the early stage, it is crucial to ensure tillering and strengthen stems. During the middle stage, it is important to strengthen the roots and ensure panicle differentiation. In the later stage, it is essential to prevent early senescence of the roots. Protect leaf growth with strong roots so as to make the rice seeds plump, and finally ensure green stems and yellow seeds. When cultivating rice, it is important to use thin water for transplanting seedlings, maintain a small amount of water for the seedlings to turn green, provide shallow water for tillering, expose the seedlings to air for better growth, keep the water level sufficient for grain-filling, provide abundant water for heading, and keep the field moist for strong seeds. One week before harvest, water should be cut off, but early dehydration should be avoided. Since the water management method of wet irrigation is adopted throughout the growth stage, weeds are easy to grow in the field, so there is an additional task of eradicating weeds. The principle of weed eradication is to "first close the whole area for weeding, then spray and kill the stems and leaves, and finally selectively kill the remaining weeds." That is to say, two to three days before transplantation, use 1,200 ml/ha of 50% propyzamide and 150 g/ha of 15% Pyrazosulfuron-ethyl to close the whole area for weeding. 10–15 days after transplantation, spray 750 ml/ha of trifluralin, cyanofluoride, and Pyrazosulfuron-ethyl on the stems and leaves, and selectively kill the stubborn weeds once more.

Compared with conventional irrigation, this irrigation method has increased rice yield by 8.56% and improved water use efficiency by 25%. The water-saving irrigation technique has the following advantages: First, it significantly reduces the transpiration rate and angle of leafing, thus reducing transpiration and improving canopy structure. Second, it significantly increases the ratio of abscisic acid to gibberellin in weak grains, the activity of sucrose phosphate synthase in stems, and the activity of sucrose synthase in grains, increasing the average grain-filling rate. Third, it significantly increases the turnover rate of non-structural carbohydrates in stem sheaths, promoting material transport and improving the harvest index.

4. Other water management technologies

(1) *Water management technology for directly seeded rice.* During the seedling stage of direct-seeding rice, the field should be kept moist. After the two-leaf-one-bud stage, shallow water should be irrigated to the field. During the tillering stage, the field should be irrigated frequently with shallow water. When the expected number of panicles is reached, the field should be dehydrated and sun-dried. Dehydration and sun-drying should be done lightly at first and then heavily, with multiple rounds to reach thorough drying. Strict control of peak seedlings should be implemented to maintain an appropriate group structure and good ventilation and light conditions, increasing the formation rate of tillers. In the middle stage, irrigation method with the focus on wet should be implemented. In the later stage, the alternative dry-wet irrigation method should be implemented. Attention should be paid to avoiding cutting off water too early.

(2) *Water management technology of tobacco-rice rotation fields.* Applying the scientific irrigation on tobacco-rice rotation fields can achieve the goal of "adjusting fertilizer with water, promoting roots with air, promoting seedlings with roots, promoting panicles with seedlings, and achieving stable and high yield." Specifically, during the period between the transplanting and regreening stage, the field should maintain a shallow water layer, and promote root growth. In the early tillering stage, a shallow water layer and wet irrigation should be combined to promote low-node tillering. When the tiller number reaches the level of three million panicles/ha or 25 days after transplantation, the field should be timely sun-dried, so as to control ineffective tillering and improve the panicle rate. During the booting and heading stages in their life, rice plants are most sensitive to the amount of physiological water supply. As they have a relatively higher demand for water in this period, irrigation should be applied timely, keeping an inch of water to promote booting; after heading, take the alternative dry-wet irrigation while using wet as the main method, and dehydrate the fields five days before harvest.

(3) *Water management technology for regenerated rice.* The key to regenerated rice water management technology is to focus on the sun-drying of first-season rice fields and the drought resistance of regenerated rice. For first-season rice, it is recommended to transplant seedlings when the water level in the field is shallow, and to irrigate alternately between dry and wet to promote tillering. In late May, the field should be dried, and seedlings should be controlled to promote the development of the root. In early June, during the stage of panicle differentiation, water should be replenished. In mid to late July, maintaining a water layer in the field is beneficial for booting and heading. During the period from late July to the harvest of early-season rice, irrigation should be alternated between dry and wet conditions. It is advisable to harvest early-season rice when the field is moist with 30% water content, where the soil is soft but is concrete enough for the harvesting machine to stand on. If regenerated rice is cultivated during the period of autumn drought, it is necessary to increase the irrigation to ensure the growth of the rice. Water management for regenerated rice mainly includes stages as follows: seven to ten days before harvesting the first-season rice, it is necessary to apply growth fertilizer and irrigate once; three to four days after harvesting the first-season rice, neither dry nor flooded field conditions are conducive to the germination of second-season rice, so it is better to keep the field moist, which can meet the water requirements for germination; after that,

alternative dry-wet irrigation should be applied in the field; in late September, a field with a water layer is beneficial for the booting and heading of second-season rice.

(4) *Water management technology for integrated rice farming.* For rice-crayfish co-cultivation water management, the water level for rice transplanting should not be too deep to prevent seedling drifting, ideally at 2–3 cm. After transplanting, it is crucial to raise the water level to 3–5 cm, ensuring that the seedlings return to green as soon as possible. After the seedlings return to green, the water level should be one inch high to promote tillering. When the total number of tillers in the field reaches 80% of the expected panicle number, the field needs to be drained and suntanned to ensure that the water level in the crayfish ditch is at least 15 cm below the field surface. The field is suntanned until the color of the rice leaves fades and the field is not muddy. If any abnormality is found in the crayfish, the water supply should be restored as soon as possible to prevent dehydration and death of the young crayfish during the field suntan. After suntan, the field should be immediately watered to a layer of 5–10 cm. During the booting stage, the water level should be increased by 10–20 cm. During the late grain-filling stage, alternative dry-wet irrigation should be mainly applied. About seven days before the rice harvest, the water level of the paddy field should be quickly decreased to about 5 cm, prompting the young crayfish and some of the parent crayfish to return from the field to the ditch. After that, water should be drained as exhausted as possible, creating a water level difference of 10 to 15 cm between the shrimp ditch and field ditch so as to facilitate sun-drying and mechanical harvesting later. During rice harvest, it is recommended to use a pulverizing harvester and leave stubble that is 40–50 cm in length. After harvesting, shallow water should be timely irrigated, prompting the rice stump regeneration buds to sprout. After that, regenerated rice becomes cheap bait for crayfish. In this process, the field should be irrigated three to four times, till water can flow between the field and the crayfish ditch. At this time, crayfish can move to the large field water region.

When chemical pesticides are used in rice-crayfish fields, it is important to control the water level. Generally, the method of deepening the field water is adopted to reduce the harm caused by the drifting of pesticide liquids to crayfish. If conditions permit, the water should be changed promptly after spraying pesticides to ensure the safety of the crayfish. Crayfish that have matured in rice fields are typically harvested and sold from late August until the end of September. Smaller crayfish can be left in the fields for further cultivation, while some are released for reproduction. The harvest of rice from rice-crayfish fields is generally planned for late September to early October.

CHAPTER 6

Group Quality Design and Cultivation for Super-High Yield of Super Hybrid Rice Group

The super-high yield of rice relies on the synergistic design between the group and individuals. Therefore, it is necessary to shape a reasonable group structure that is highly valued by the academic community and production practice. To achieve high yields in super rice, it is crucial to effectively manage the balance between group quantity and quality during cultivation. Therefore, it is required to figure out the main growth features of super hybrid rice and the formation law of rice yield components. Based on this, we also need to identify quality indicators for the group, develop relevant key technologies for group management, implement group quality design and cultivation, and ultimately achieve the goal of achieving super-high yields.

I. Classification of Main Growth Stages of Rice

Conventionally, the process from the germination of rice seeds to the formation of new seeds is referred to as the life of rice. According to the different periods of rice morphology, physiology, and other characteristics, the whole life cycle of rice can be divided into two growth stages: nutrient growth stage and reproductive growth stage. The nutrient growth stage is a period from seed germination to panicle differentiation; in this stage, organs such as roots, stems, leaves, and tillers are formed. The reproductive growth stage is the period from panicle differentiation to maturity and harvest. This stage mainly includes booting, heading, fruiting, and new seed formation. The full growth period of rice ranges from less than 100 days to more than 180 days, in which the reproductive growth period is generally 60–70 days, and the rest is the nutrient growth period. Due to different nutrient growth periods, different rice varieties have different growth periods. In rice production practice, apart from direct-seeded rice, the process of seedling cultivation and transplanting of rice is divided into two stages: seedling bed stage and paddy field stage. The nutrient stage of rice is divided into the seedling bed nutrient stage and the paddy field nutrient stage. The seedling bed vegetative stage can be divided into three stages: the germinating stage from seed germination to incomplete leaf expansion, the seedling stage from incomplete leaf expansion to the full emergence of the third leaf, and the matured seedling stage from the emergence of the fourth leaf to transplantation (if the nursing seedling cultivation method is adopted, the seedlings are transplanted into the paddy field directly at the three-leaf stage, therefore there is no matured seedling stage).

TABLE 6.1 Division of main growth stages of rice

Major growth stage	Location of growth	Detailed growth stage	Explanation	Corresponding growth period in field
Nutritional growth	Seedling field	Sprouting stage	From seed germination to incomplete leaf expansion.	Early growth stage
		Seedling stage	From incomplete leaf expansion to full emergence of the third leaf.	
		Tillering stage	From the emergence of the fourth leaf to transplanting.	
	Main field	Recovery stage	Transplanted rice's leaves turn green, and new leaves start to grow normally again.	
		Effective tillering	Effective tillering and ineffective tillering stages.	
Reproductive growth	Main field	Young panicle formation	From the start of ear differentiation to a certain stage before heading.	Mid-growth period
		Flowering to maturity	From heading to ripening.	Late growth period

The paddy field nutrient growth stage can be divided into the regreening stage and the tillering stage. During the regreening stage, which lasts for one to seven days, rice plants should be transplanted from the seedling field to the paddy field, with the transplanted rice leaves turning green and new leaves returning to normal growth. The tillering stage is divided into the effective tillering period and the ineffective tillering period. The effective tillering period begins after the regreening and continues until the number of stems in the paddy field reaches the planned number of panicles. After this, rice enters the ineffective tillering period.

As rice continues to grow, it begins to enter the reproductive growth stage, which can be subdivided into the panicle development stage and the flowering and grain-setting stage. The panicle development stage is specifically the period from panicle differentiation to heading, including the reproductive organ differentiation period from the initiation of panicles to the point when the flag leaf tip emerges and the reproductive cell formation period from the emergence of the flag leaf to just before heading. The flowering and grain-setting stage refers to the period from the emergence of panicles to maturity, which can be divided into the flowering period from the beginning of the panicle's leaf sheath to the end of the flowering and pollination, and the grain-setting and maturing period from the end of the pollination to the maturity of the harvest.

To facilitate rice production, the paddy field growth period can be divided into three stages according to the cultivation management process, namely, the early growth stage from transplanting seedlings to panicle differentiation, the middle growth stage from panicle differentiation to heading, and the late growth stage from heading to maturity and harvest.

Because of the large panicles and grains, the super hybrid rice has a longer grain-filling period, about 10 days longer than the ordinary mid-season or late-season rice, and some even up to 45 days or more. Therefore, during the grain-filling stage of super hybrid rice, special attention should be paid to water management of alternative dry-wet irrigation or continuous wet irrigation. Dehydration should not occur too early. At the same time, the grain-filling stage should be scheduled in a suitable climate.

II. Time Determination of Rice Yield Components

Yield in rice production refers to the economic yield, namely the yield required for the cultivation purpose of grain production. Super hybrid rice boasts the characteristic of large biomass. Currently, the economic coefficient of super hybrid rice varieties bred in China can reach about 50%, with equal amounts of grain and straw. Therefore, properly increasing the total biomass of super hybrid rice is one of the important ways to increase rice yield.

Rice grain yield in production is composed of four elements: the number of effective panicles per unit area, the number of grains per panicle, grain-setting rate, and grain weight (thousand-grain weight). The relationship between them can serve as the basis for the design of high-yield group quality.

Yield (kg/ha) = Effective panicles per unit area (ha) × Grains per panicle × Grain-setting rate (%) × Thousand-grain weight (g) × 10^{-6} .

The formation process of each component of rice yield is also the process of organ formation in the growth and development of rice. The formation of each component corresponds to a certain time in the development of rice.

TABLE 6.2 The formation of the four factors affecting rice yield corresponds to each growth stage

<table>
<tr><td rowspan="5">Nutritional Growth Stage</td><td>Seedling stage</td><td rowspan="2">Seedling field stage</td><td rowspan="2">Determining panicle number</td></tr>
<tr><td>Seedling tillering stage</td></tr>
<tr><td rowspan="3">Tillering stage</td><td>Regreening stage</td><td rowspan="3">Determining panicle number and grain number</td></tr>
<tr><td>Effective tillering</td></tr>
<tr><td>Ineffective tillering</td></tr>
<tr><td rowspan="6">Reproductive Growth Stage</td><td rowspan="3">Booting stage</td><td>Differentiation</td><td rowspan="3">Determining panicle number, grain number and grain weight</td></tr>
<tr><td>Formation</td></tr>
<tr><td>Maturity</td></tr>
<tr><td rowspan="3">Flowering and maturing stage</td><td>Milk ripening</td><td rowspan="3">Determining grain weight</td></tr>
<tr><td>Wax ripening</td></tr>
<tr><td>Maturing</td></tr>
</table>

The four elements of output are interrelated and mutually constrained. Therefore, in order to achieve high yields, it is necessary to regulate the four elements of output throughout the entire growth period based on the characteristics of different varieties and achieve a reasonable group composition. For example, the number of effective panicles per unit area increases with the increase of basic seedlings within a certain range. However, when the number of effective panicles per unit area exceeds a certain range, the contradiction between panicle number and grain number increases; that is, an excessive number of effective panicles per unit area will lead to a decrease in grain number per panicle. When the increase in the number of panicles cannot compensate for the decrease in the number of grains per panicle, it will lead to a decrease in yield. Therefore, only by selecting suitable varieties, strengthening cultivation management, coordinating the relationship between individuals and groups, and adjusting the optimal composition of various yielding factors can we achieve super-high yields.

1. The formation stage of panicles

The characteristic of super hybrid rice is its large panicle shape, which leads to super-high yield. However, when the number of effective panicles per unit area exceeds a certain range, it will increase the contradiction between panicle number and grain number, resulting in an insufficient total spikelet number required for high yields. Therefore, it is not advisable to plant rice too densely. The focus should be on large panicles based on a stable number of panicles.

The number of rice panicles is affected from the tillering stage, even from the seedling stage, and this influence lasts until seven to ten days after the peak tillering stage. Among them, the main period determining the number of panicles is the peak tillering stage. During the tillering stage, the most important thing is the coordination between the quantity and quality of panicles.

The number of panicles per unit area is influenced by the number of plants, the number of tillers per plant, and the panicle formation rate. The density of planting and the survival rate after transplanting determine the number of plants per unit area. The number of tillers per plant and the survival rate of tillering are closely related to the strength of seedlings. Therefore, the determination of the number of panicles is in the tillering period, and the foundation is in the seedling stage. Sparse planting is beneficial for the formation of large panicles, which promotes ventilation and light penetration in the field, reduces the harm of pests and diseases, and helps to form a strong root system. Super hybrid rice has strong individual characteristics. Only by providing sufficient space for individual development can we maximize the advantages of individuals and tap into their high-yielding potential. Therefore, the cultivation of super hybrid rice with high yield should shift from the previous "reasonable dense planting" to "reasonable sparse planting." Reasonable sparse planting should be determined by the variety and geographical conditions. Mid-season rice with a longer growth period should be sparsely planted, while late-season rice with a shorter growth period should be densely planted. Rice with strong tillering capability should be planted sparsely, while rice with weak tillering ability should be planted densely. Fertile paddy fields should be planted sparsely, while lean fields should be planted densely.

The change in the number of rice groups reflects the situation of tillering and booting and significantly affects the yield. Reasonable dynamics of tiller number and high panicle formation rate are among the basic characteristics of super-high-yielding groups. The reasonable stem and tiller dynamic model for super hybrid rice should be as follows: during the transplanting period, determine the primary seedlings and the critical leaf age of effective tillering (total number of leaves N – number of elongated internodes n) reaches enough seedlings for high yield. The leaf age of the stem elongation stage (reciprocal of n – 2) is the peak seedling period, with the maximum number of seedlings being 1.2 to 1.3 times the number of effective panicles. The heading period should reach the appropriate number of panicles.

Main cultivation measures to increase the number of panicles:

1) *Cultivate strong seedlings.* It provides a good foundation for early greening and early tillering after transplantation. Taking water nursery as an example, the recommended seeding density for super-high-yielding sparse seedling cultivation is 60% to 80% of the recommended seeding density for rice seed commercial bags. Around 5 kg of 46% urea is applied per *mu* during the one-leaf-one-bud stage as weaning fertilizer, and around 5 kg of 46% urea is applied per *mu* three to five days before transplanting as transplanting fertilizer. During the seedling stage, plant growth regulators can be applied to cultivate strong seedlings, promote dwarfing of seedlings with tillering, increase chlorophyll content, promote root growth, and improve seedling resistance. Common plant growth regulators include propiconazole, paclobutrazol, and others such as Nengbawang, S-inducing agents, and Tianda 2116, all of

which can achieve the effect of controlling height and promoting tillering. Cultivate strong seedlings with two to four tillers before transplanting to substitute tiller for seedlings and save the seeds.

2) *Reasonable sparse planting.* Based on the planned harvest panicle number and the evaluation of the ability to retain water after transplanting seedlings, determine the appropriate number of seedlings to be transplanted. For transplanting rice, it is advisable to use wide rows and narrow spacing. Due to different varieties of hybrid rice, the planting density may also vary. The recommended planting density is 10,000 to 12,500 holes per *mu*. Taking Y Liangyou 900 as an example, the recommended row and plant spacing is 30 cm × 20 cm or 26.7 cm × 20 cm, with a yield of 900 kg per 900 *mu*. Different planting densities should be treated differently according to the soil fertility level. For rice fields with higher fertility levels, a row spacing of 30 cm and a plant spacing of 20–30 cm is recommended. For rice fields with moderate fertility levels, a row spacing of 26.7 cm and a plant spacing of 20–26.7 cm is recommended.
3) *Promote early tillering at lower nodes and suppress excessive tillering at higher nodes.* Transplanting should be carried out at an early stage when the rice plant has four leaves and one tiller. The seedling age should be limited to 30 days during water seedling cultivation to ensure a sufficient effective tillering period. After transplanting rice, you can apply physical and chemical products that encourage root regrowth such as Nengbawang, S-inducing agents, Yizailing, and Dadichun to promote early development of low-position tillers.

Fertilization is an important measure to promote tillering, so sufficient base fertilizer and early tillering fertilization should be applied. The base fertilizer accounts for about 40% of the total fertilizer applied, with 100 kg of organic fertilizer (5% of the total nitrogen, phosphorus, and potassium) and 50 kg of compound fertilizer (45%, 15-15-15) per *mu*. Tillering fertilizer accounts for about 20% of the total fertilization. Tillering fertilizer should be applied with quick-acting nitrogen fertilizer. The first application should be done five to seven days after transplanting, with about 8 kg of urea per *mu* and about 8 kg of compound fertilizer (15-15-15). The second application should be made 14 days after transplanting, with 3–5 kg of urea and 7.5 kg of potassium fertilizer per *mu* to promote the early growth of seedlings. After the number of tillers is sufficient, the field should be exposed and dried in a timely manner to control the ineffective tillers at high nodes, so as to concentrate the nutrient on the early emerging effective tillers and improve the tillering rate.

2. Formation stage of rice grain

The number of rice grains is determined during the period of panicle differentiation. The complete formation of the panicle's morphology and internal reproductive cells is a continuous process. Ding Ying divides rice grain development into eight stages (TABLE 6.3). Among them, the first four stages are the panicle formation periods (reproductive organ formation periods), and the last four stages are the booting periods (reproductive cell development periods). The time of rice panicle development varies due to different varieties (combinations) of growth period, temperature, and nutrition.

The entire process takes 25 to 35 days. The growth period of super high-yielding rice is longer, and the time required for panicle initiation is longer than that of conventional rice, which is 30 to 35 days.

TABLE 6.3 Eight stages of rice grain development

Stage	Explanation	Method for identifying spikelet differentiation stage					Days after stage	
		Naked eye	Leaf age index (%)	Leaf age remain-ing (%)	Spikelet growth length (cm)	Time Duration (day)	Days to Heading (Y Liangyou 900)	Days to Heading (Xiang-liangyou 900)
1st	First primary branch differentiation stage	Invisible	76 ±	3.0 ±	<0.1	2–3	33.5	34
2nd	Secondary branch differentiation stage	Hair emergence	82 ±	2.5 ±	0.1–1	4–5	28.5	29
3rd	Secondary branch differentiation stage to heading	Hair expansion	85 ±	2.0 ±	1–2	6–7	22	22.5
4th	Male sterile spikelets visible stage	Grain visible	92 ±	1.2 ±	4	4–5	17.5	17.5
5th	Flower mother cell formation Stage	Grain fully enclosed	95 ±	0.6 ±	25	2–3	13.5	14
6th	Flower mother cell meiosis stage	Grain half visible	97 ±	0.5 ±	60	2	10	10.5
7th	Flower pollen development stage	Full grain visible	100 ±	0	90	7–8	6.5	7
8th	Flower pollen content fully visible stage	Rapid heading	—	—	—	2–3	3	3

The number of grains per panicle mainly depends on the number of differentiated florets and degenerated florets in the young panicle. That is to say, spikelet number per panicle equals to the number of differentiated panicles minus the number of degenerated spikelets. Therefore, we should try to increase the initiation of the spikelets and reduce the degenerated spikelets as much as possible. The number of differentiated tillers is positively correlated with the stem thickness and nutrient content of the plants, so the tillering stage before initiation is the period that determines the number of rice grains. The large size of the panicles of super hybrid rice is related to the higher number of secondary branches, so it is important to focus on increasing the initiation of secondary branches during cultivation.

Cultivation measures to increase the number of grains per panicle are as follows:

1) *Ensure stalks are stout before panicle initiation.* During the tillering stage in paddy fields, to ensure the necessary nutrient supply for rice plant growth, it is necessary to make the rice stems thick, increasing the number of tillers and forming larger panicles.
2) *Provide sufficient nutrients for panicle differentiation.* Panicle fertilizer is the prerequisite for large panicles and a high grain rate, accounting for about 40% of the total fertilizer application.

   Usually, when the initiation of spikelets begins and the leaf color is pale, it is necessary to apply additional flower-promoting fertilizer (generally during the second stage of spikelet initiation on the main stem, with nitrogen fertilizer as the main component) to form more florets. In the middle stage of young panicle differentiation and development (about 18 days before booting), if the leaf color is lighter, extra flower-preserving fertilizer (applied in the fourth stage of young panicle differentiation of the main stem, N, P, and K are combined to control N) should be applied to reduce spikelet degeneration and improve the breeding rate of spikelets. The flower-promoting fertilizer with a high yield of 900 kg/*mu* is applied at a rate of about 10 kg of urea per *mu*, about 10 kg of compound fertilizer (15-15-15) with 45% nutrient content, and about 10 kg of chloride fertilizer per *mu*; the flower-preserving fertilizer is applied at a rate of about 5 kg of urea per *mu* and about 8 kg of chloride fertilizer per *mu*.

   It is important to note that the key measure for achieving a yield of over 900 kg/*mu* is the proper application of panicle fertilizer after nitrogen application and during the appropriate period. The rate of panicle fertilizer used during super-high-yielding rice cultivation is 40%, which is the biggest difference between it and ordinary rice cultivation.

   In addition, in the cultivation of super-high-yielding crops, the optimal period for applying panicle fertilizer should not be the traditional fourth-to-last leaf (half-leaf-stage before the initiation of the main stem panicle) and second-to-last leaf (around the third stage of the initiation of the main stem panicle), because applying panicle fertilizer during these two periods is not conducive to the formation of large panicles and increases the length between the first and second internodes at the base, which is not conducive to lodging resistance. Panicle fertilizer should be applied during the second and fourth stages of the initiation of the main stem panicle, so as to achieve the dual effects of promoting large panicles and lodging resistance.
3) *Increase the accumulation of dry matter before rice booting.* Before booting, make sure to provide the necessary nutrients for growth, but avoid overstimulating growth. Accumulate photosynthetic products to ensure that more substances are transferred to the grains after booting, which will increase the rate of grain-setting.

3. The grain-setting formation stage

The grain-filling rate of rice is a comprehensive reflection of the combination of storage, source, and flow, and it has the largest variation among the four yielding factors. The characteristics of super hybrid rice are large panicles, double grain-filling, and a high grain-setting rate, which is also the prerequisite for achieving high yields. However, the grain-filling rate is very sensitive to environmental conditions, and it can be affected if the ecological and cultivation conditions are slightly unsuitable. Therefore, attention must be paid to all aspects.

The time that affects the grain-setting rate is relatively long, which is from the initiation of rice panicles to maturity. Among them, the most sensitive periods are the period of floral initiation, reduction division phase, booting and flowering, and grain-filling stage. That is, 20 days before and after booting (a total of 40 days) are the critical periods that affect the grain-setting rate.

Various factors influence the grain-setting rate. Poor seedling quality, excessive plant density, premature field closure, unfavorable field climate conditions, inadequate mid-season sunlight, deficiency of nutrients and water during meiosis, unfavorable weather during flowering, inadequate nutrition, and insufficient water and fertilizer during grain-filling all directly contribute to a decrease in the grain-setting rate.

Cultivation measures to increase grain-filling rate are as follows:

1) *Research should be conducted to maintain the optimal group structure of rice plants and grain clusters according to different geographical conditions.* Specific methods include cultivating strong seedlings with tillers and well-developed root systems in the early stage, timely transplanting and reasonable thinning, baking the field in the middle stage to increase starch accumulation, etc.
2) *Arrange the most suitable season for booting to prevent the influence of high and low temperatures.* During the flowering stage of booting, if there is dry and cold weather (such as frost wind), irrigation can be used to moisturize and warm the plants, reducing the damage caused by low temperature and dryness and improving the grain-setting rate.
3) *Apply grain fertilizer as topdressing in the later stage.* Super hybrid rice has large grains, but if the management of fertilizer and water is unsuitable, it is prone to premature senescence and nutrient deficiency in the later stage. Therefore, in the early stage of grain-filling, urea 50 kg, potassium dihydrogen phosphate, and foliar fertilizers such as spray fertilizer should be mixed with water and sprayed on the leaves of rice per *mu*, reducing the percentage of empty grains, and increasing the grain-filling rate and grain weight.
4) *Late-stage supplementary fertilizers.* Spraying silicon fertilizer on the leaves can not only improve the grain-setting rate but also reduce heat damage and slow down its decline. Plant growth regulators such as Baiwang, S-inducing agents, Tianda 2116, etc. can prevent cold damage, improve stress resistance, and increase chlorophyll content, thereby improving the grain-setting rate.
5) *Optimize water management.* The milk ripening and grain-filling period is a critical period for super-high-yielding grain-filling, and alternative wet-dry water management should be applied. If there is a drought and lack of water, it will cause premature leaf senescence,

decrease in photosynthetic efficiency, and affect grain-filling. If waterlogged irrigation is applied, the reduction of oxygen in the soil will affect the later vitality of the roots and cause premature leaf senescence, resulting in poor grain-filling. Therefore, during the milk ripening period of rice, it is essential to alternate between wet and dry conditions, maintaining soil moisture without waterlogging and ensuring irrigation without water deficiency, which facilitates grain-filling. After the ripening stage, the water requirement of rice significantly decreases. At this time, the field should not be waterlogged to promote maturity.

4. The formation stage of grain weight

Due to the larger grain size of hybrid rice, the impact of grain weight on yield is greater than that of conventional rice, especially for some short-growth-period combinations (varieties). Since the high yield of super hybrid rice is often achieved through the advantage of large grains, the weaker advantage of large panicles in this case may result in a poor performance of the varieties in rice production. Therefore, promoting grain weight should be emphasized in cultivation.

The grain weight is determined by the size of the husk and the degree of grain-filling. The size of the husk is determined during the period of panicle differentiation and development. About 20% of the grain-filling material comes from the storage material before booting. Therefore, the period of panicle differentiation and development is an important period for establishing a grain weight foundation. The spikelet of rice grows most vigorously in the meiosis stage, and its influence on grain weight is also the greatest. Therefore, this period is often referred to as the first determining period of grain weight. After booting in rice, when the size of the husk is fixed, the degree of endosperm filling determines the volume and weight of the grain. Therefore, the peak period of grain-filling after booting is called the second determination period of grain weight.

Super hybrid rice has large panicles, more secondary branches, and weak flowers, resulting in uneven grain weight. Factors affecting the rate of rice grain-filling include the position of spikelet initiation, the group structure of rice plants, the rate of root and leaf senescence, soil nutrients and moisture, and climate. When the temperature is too high or too low, the synthesis of plant photosynthetic products decreases, the filling rate slows down, and the grain weight decreases. When the group structure is too shaded, the insufficient reserve of carbohydrates before flowering and excessive spikelets per unit area are not conducive to the increase in grain weight.

Measures to increase grain weight are as follows:

1) *Ensure good nutritional conditions.* Good nutritional conditions during the late stage of young panicle development can help the husks grow larger to accommodate more grain-filling substances.
2) *Ensure grain-filling and grain-setting efficiency.* After booting, when the size of the husk has been fixed, the cultivation measures for the second determination period of grain weight, which coincides with the grain-filling and grain-setting period after booting, are basically the same. This includes the application of grain fertilizer and the application of chemical

products in the later stage, which delays leaf senescence, increases photosynthetic capacity, accelerates grain-filling speed, increases grain weight, and improves quality. A strong emphasis on water management is necessary, particularly during the grain-filling and ripening period. Implement moist irrigation to nurture root development and preserve foliage, promoting the accumulation of ample photosynthetic products and ensuring good grain-filling.

III. Group Quality Design for Super-High Yield of Super Hybrid Rice

The yield of super hybrid rice depends on four factors: the number of effective panicles per unit area, the number of grains per panicle, the grain-setting rate, and the grain weight (thousand-grain weight). Different varieties have different yield compositions due to differences in panicle shape and grain weight. To achieve high yields in super hybrid rice, growth and development indicators should be designed based on the characteristics of each variety so that corresponding segmented target dynamic management measures can be taken to organically coordinate and unify the four elements of yield composition. Currently, to achieve a target yield of 900 kg/*mu*, it is necessary to use large panicles to obtain high yields based on the determined number of panicles. At present, the super-high-yielding group indicators of the main super hybrid rice varieties in China are shown in TABLE 6.4, which can be used as reference for the design of similar varieties, but each variety has certain differences in different growing seasons and environments and should be treated differently.

TABLE 6.4 Yield design of different super hybrid rice varieties

Variety	Y Liangyou 2	Y Liangyou 900	Xiangliangyou 900	Yongyou 12
Density (per ten-thousand holes/*mu*)	1.11	1.11	1.11	1.11
Effective tillers (per ten-thousand panicles/*mu*)	20	16	15	15
Total grains	200	250	270	300
Total spikelets (per ten-thousand/*mu*)	4000	4000	4050	4500
Grain-setting rate (%)	88	88	88	88
Thousand-grain weight (g)	26.5	26.5	26.5	23.5
Target yield (kg/*mu*)	932.8	932.8	944.5	930.6

1. General ideas on the group quality design of super-high-yielding super hybrid rice

1) Expand the total group glumes to construct a highly photosynthesis-efficient and accumulating group at the booting and grain-setting stage.
2) Through growth diagnosis at different leaf ages, appropriate measures are taken to regulate the growth of various organs and the development of the group in a targeted and quantitative manner.
3) Follow the path of "precise seedling establishment to ensure early growth-controlling tillers while promoting medium tillers—prioritizing strong panicles in the later stage" in super-high-yielding cultivation. At first, it is necessary to compress the group structure and the total growth number, and then build a rational group structure based on the healthy individuals.
4) Adopt techniques such as fertilization and irrigation to promote effective and efficient growth, and control ineffective and inefficient growth.
5) To achieve maximum economic and ecological benefits with minimal input, precise quantification techniques are employed while ensuring high-yielding and high-quality rice populations.

2. Key technical indicators for group quality design

1) *Yield design index.* Make sure effective panicles per *mu* of more than 150,000 panicles, total glumes of more than 40 million, a grain-setting rate is more than 85%, and thousand-grain weight equals 26–28 g.
2) *Precise seeding.* Sow seeds at the right time, sow sparsely, and cultivate age-appropriate strong seedlings with leaf tillers.
3) *Reasonable sparse planting.* Rice should be transplanted early, and the spacing between plants should be wider in rows and narrower between plants. The recommended planting density is 10,000 to 12,500 plants per *mu*. It is necessary to extend the effective tillering growth period and increase the growth amount to form large panicles, with a panicle formation rate of over 75%, a leaf area index of 7.5–8 during the booting stage, and 4–5 green leaves during the maturity stage.
4) *Promote early and rapid growth of rice.* Make the number of seedlings enough, timely sun-drying and control seedlings, and shape strong stalks and large panicles.
5) *Precise application of fertilizer.* Organic fertilizer and inorganic fertilizer are used concurrently; the amount of organic fertilizer accounts for 20% to 30% of the total nitrogen, nitrogen, phosphorus, potassium, silicon, and other fertilizers are applied in conjunction. Precise application of nitrogen fertilizer, base tiller fertilizer: panicle fertilizer = (6:4) – (5:5). Reduce the amount of nitrogen fertilizer in the first and middle stages to reduce ineffective

tillering (control the ineffective composition of the population). Increase the proportion of panicle fertilizer application to promote large panicles and improve the photosynthetic function of canopy leaves. Panicle fertilizer is applied twice at the second and fourth stages of young panicle initiation in the main stem, considering the attack on large panicles, lodging resistance, and high grain-setting rate at the same time.

6) *Scientific irrigation.* Adopt the irrigation technology of "shallow water–sun-drying–wet" during the whole life cycle and start alternative dry-wet irrigation during the grain-setting period to improve the vigor of the root system (to feed the roots and protect the leaves to improve the fruiting rate), and do not cut off the water too early.

CHAPTER 7

High-Yielding and High-Efficiency Cultivation Technology Mode of Super Hybrid Rice

I. Cultivation Mode of Super Hybrid Rice with a High Yield of 900 Kg/*Mu*

1. General information

The base is located at an altitude of 300–500 m, with a mild climate, abundant light, and heat, plentiful rainfall, an average annual temperature of 16.0°C, annual rainfall of about 1,300 mm, and a frost-free period of about 270 days. Soil organic matter content is about 35 g/kg, with a pH ranging from 5.5 to 6.5. The soil layer is deep, with strong water and nutrient retention capacities, complete drainage and irrigation facilities, and good ecological conditions. According to the yield requirements, the panicle structure is designed as follows: approximately 2.4 million panicles per ha, with an average of 280 grains per panicle, a fertility rate of 90%, and a thousand-grain weight of around

27 g. High-yielding cultivation techniques revolve around how to improve the above yield-contributing factors through corresponding water and fertilizer management.

2. Seedling raising and transplanting

In mid-April, soaking of seeds generally begins. Before soaking, the seeds should be sun-dried, selected, and disinfected. Creating favorable conditions for germination ensures that the rice seeds germinate quickly, uniformly, and vigorously. Germinating seeds should be cultivated in logged water. The paddy field for seedling cultivation should be selected with a sunny and wind-sheltered location, convenient for irrigation and drainage, and with a deep and fertile topsoil. After plowing the paddy field, apply 600 kg of 45% (15-15-15) compound fertilizer per ha, and 112.5 kg of potassium chloride as base fertilizer. After two to three days, level the seedbed and sow the seeds. When the seedlings have two leaves and one bud, apply weaning fertilizer with 60 kg of urea per ha to ensure that the seedlings can smoothly transition from self-sufficiency to adapt to new environments. Three to four days before transplanting, apply 105 kg of urea per ha as transplanting fertilizer to prepare for early and vigorous growth in the paddy field after transplantation. During the seedling stage, it is necessary to prevent pests. One day before transplanting, a long-lasting pesticide should be sprayed to make the seedlings carry the pesticide into the field. This effectively reduces pests and diseases in the early stages.

Before transplanting, the field should be finely tilled. Use a medium-sized tilling machine to plow the paddy field to a depth of 20–25 cm, then level it with a harrow. Ensure that a 3-cm-deep water layer is irrigated without mud. According to the yield index, the planting specification adopts wide and narrow rows, with a width of 40 cm for wide rows and 23.3 cm for narrow rows, with a spacing of 20 cm between plants, that is, 158,000 seedlings per ha. The wide rows are set in the east-west direction, and a special row marker is used for row marking and transplanting, which is beneficial for ventilation and light transmission in the middle and later stages. When transplanting, rice seedlings should be shallowly inserted, with the tillering node inserted into the mud 2–3 cm deep. Shallow insertion is the premise for low-position tillering to occur early and in large quantities, ensuring sufficient rice seedlings and large panicles. At the same time, it is important to avoid excessive or missed insertion. Around four days after transplantation, if there are any dead rice seedlings, they should be promptly replaced.

3. Apply fertilizer reasonably

1) *Target yield.* The target yield is 3.5 tons/ha, N requirement per ton of rice: 18 kg, total N requirement $=13.5 \times 243$ kg/ha of pure N. Soil supply of pure nitrogen: 120 kg/ha, then pure nitrogen to be applied: $243 - 120 = 23$ kg/ha. Fertilizer utilization rate: 40%, then the need for supplementary nitrogen per ha: $123 \div 40\% = 307.5$ kg/ha. The fertilizer ratio before and after fertilization is base tiller fertilizer: panicle fertilizer = 6:4, then the base tiller fertilizer

requires pure nitrogen 184.5 kg/ha, and the panicle fertilizer requires pure nitrogen 123 kg/ha. To achieve high-yielding cultivation, it is important to use phosphorus, nitrogen, and potassium fertilizers in a balanced and reasonable manner. The ratio of nitrogen, phosphorus, and potassium should generally be 1:0.6:1.1. Fertilizers should be applied in three categories: base fertilizer, panicle fertilizer, and tillering fertilizer.

2) *Base fertilizer.* After plowing and harrowing the field twice, base fertilizer is applied in two separate applications. For the first time, apply 450 kg of 45% (15-15-15) compound fertilizer per ha and mix well. Then, during the first plowing, base fertilizer is deeply applied; during the second plowing, 750 kg of carbendazim per ha and 450 kg of compound fertilizer (45%,15-15-15) are evenly mixed and applied. Then, apply sprinkle base fertilizer during the second plowing of the field.
3) *Tillering fertilizer.* For the first time, 6 days after transplanting, apply 105 kg/ha of urea and 90 kg of potassium chloride (containing 60% K_2O) when turning green; for the second time, 13 days after transplanting, apply 40 kg/ha of urea/ha, 112.5 kg of potassium chloride. At the same time, during this period, pay attention to the use of warm irrigation and open field ventilation to promote the growth of a strong and developed root system, which is conducive to improving the absorption capacity of the root system and promoting early and rapid tillering.
4) *Panicle fertilizer.* Divide the application into three stages: panicle initiation fertilizer, flower-promoting fertilizer, and grain fertilizer. When the main stem enters the second stage of panicle initiation, apply 90 kg of urea per ha, 225 kg of compound fertilizer (45%), and 112.5 kg of potassium chloride to promote branch initiation and lay the foundation for the formation of large panicles. When the main stem enters the fourth stage of panicle initiation, apply 45 kg of urea/ha and 112.5 kg of potassium chloride to provide sufficient nutrients for the panicles, prevent the degeneration of differentiated spikelets, and ensure the formation of final large panicles.
5) *Grain fertilizer.* During the booting stage, spray 15 bags of grain mixed with 750 kg of water per ha on the leaves. Five days after booting, sprinkle 15–30 kg of urea per ha to reduce empty grains and increase the grain-setting rate and grain weight.

4. Scientific water management

Plow the field and dig a ditch around the big field with a depth of 30 cm and a width of 20 cm; dig cross ditches on the paddy field with the same depth and width as the ditch. In addition, for every 300 m^2 or more, dig a row ditch 20 cm wide and 20 cm deep to facilitate drainage irrigation and field-drying. After transplanting, irrigate with shallow water, and during the period from regreening to the critical stage of effective tillering, practice intermittent irrigation. This means watering with one to two cm of water, allowing it to naturally dry for three to four days before watering again with one to two cm of water, repeating this process periodically. When 80% of the expected number of seedlings (2.55 million per ha) is reached (2.04 million per ha), drainage and sunshine will be started, and the method of multiple light sunshine will be adopted until the color of the leaves turns pale. If it reaches the time that panicle fertilizer should be applied, but the leaves have not turned pale, the sun-drying should continue without irrigation and fertilizer. In the grain-filling and maturing period, the intermittent irrigation and alternative dry-wet water management should be applied. Apply wet irrigation in the flowering period and dry in the later period to ensure root vigor, prevent early failure, and improve the grain-setting rate and grain-filling rate. Effectively implement the alternation between dry and wet conditions, maintaining clear water and hard soil to nurture the roots, preserve the leaves, and increase grain weight. This approach aims to achieve the required effective number of panicles, ensuring vitality from youth to maturity and ultimately obtaining high yields.

5. Integrated prevention and control of diseases and pests

Focusing on prevention, disease, and pest control, adopt an integrated approach that combines prevention and control methods. The key varieties are rice thrip during the seedling stage and rice stem borer, rice leaf roller, rice planthopper, rice blast, and sheath blight during the field stage.

It is recommended to use the following pesticides for control:

1) *Rice thrip, rice stem borers, and leaf rollers.* Use chlorantraniliprole (Conke), diafenthiuron, profenofos, chlorpyrifos, abamectin, flonicamid, etc.
2) *Rice planthopper.* Use chlorpyrifos, thiamethoxam, imidacloprid, chlorfenapyr (Kangkuan), fenitrothion, dimethoate, dichlorvos, etc.
3) *Rice blast (mainly for prevention).*

    a) Rice blast. To prevent rice blast, use triadimefon 15–20 days after transplanting in combination with insect control.
    b) Stem blast. Spray Fuji No. 1 one to three days after breaking the ground; spray ethyl garlicin after booting. If stem blast symptoms are found, use 80 ml of 6% spring thunder mycelium per *mu* to treat the disease.
    c) In areas where blast frequently occurs and humidity is high, it is recommended to add a small amount of triazole to the pesticide spray as a preventive measure.

4) *Sheath blight.* Use amistar, paclobutrazol, glyphosate, jinggangmycin, etc.
5) *False smut.* Use jinggangmycin, triadimefon, mancozeb wettable powder, etc. Apply medicine in a timely manner after the rice booting stage (five to seven days before rice bursting). If a second prevention is needed, the medicine should be applied during the rice booting stage (around 50% booting), as the effect of prevention during the panicle formation stage is poor.

II. Cultivation Mode of Super Hybrid Rice with a High Yield of 1,000 Kg/*Mu*

1. General information

The base is located in fields with gentle terrain, fertile soil, a deep soil layer, strong water and fertilizer retention capacity, complete drainage and irrigation facilities, and good ecological conditions. Located at an altitude of 300 to 550 m, mild climate, abundant light and heat, abundant rainfall, annual frost-free period of 230 days or more, annual ≥10°C active cumulative temperature of 4,800°C, the total annual solar radiation is greater than 100 kcal/cm^2, the annual sunshine hours of more than 1,400 hours, the annual rainfall of about 1,350 mm. Distinguished from the 900 kg/*mu* model, the design of the panicle-grain structure of this model should be: the number of effective panicles per ha is 2.5–2.7 million panicles, the total number of grains per panicle is 280–330 grains, the grain-setting rate is more than 90%, and the thousand-grain weight is 26–28 g. In cultivation management, emphasis is placed on key techniques such as "timely and precise seeding for robust seedling growth," "optimal spacing to promote lower node and larger panicle development," "balanced fertilization and stem reinforcement to prevent lodging throughout the entire growth period," "moist and well-aerated irrigation," and "early prediction and prevention of diseases and pests, employing integrated pest management."

2. Cultivate strong seedlings with many tillers

Adopt germination of seed-soaking and implement waterlogged seedlings. Use 15 kg/ha of seed in the field, and suntan the seed for one to two days before soaking. Choose rice fields with good irrigation and drainage system, adequate sunlight, fertile soil, and easy seedling emergence (avoiding fields with deep mud or cold water immersion) and use them as seedbeds for transplanting seedlings to the main field nearby. Seed-sowing is adopted in mid-April. Before sowing, the seeds should be disinfected with strong chlorine essence and then soaked in water to germinate till the seed sprouts grow to half the size of the grain. Each ha of the seeding field needs 112.5 kg of seeds so as to cultivate strong seedlings with numerous tillers.

The paddy field needs to be windward and sunny, convenient for drainage, with deep and fertile soil. After plowing the paddy field, apply base fertilizer. Apply 600 kg of 45% (15-15-15) compound fertilizer per ha as base fertilizer for the paddy field. After two to three days, level the seedbed and sow the seeds. After sowing, insert a 180-cm-long bamboo piece every one m on the seedbed as a low arch, then cover it with plastic film. Press the edges of the film firmly with soil, seal it to keep it warm, and prevent it from being blown open. Before the seedlings turn green, seal and insulate them to promote root growth and emergence. After the seedlings turn green, they need to be ventilated and refined to prevent seedling damage from high temperatures, water shortage, and disease. After rice reaches the two-leaf-one-bud stage, uncover the film in good weather, and spray pesticides such as Yizailing and 0.5% urea solution after uncovering the film to prevent diseases and promote vigorous seedling growth. Three days before transplanting rice, 75 kg/ha of urea should be applied as bridal fertilizer in the seedling field. During the seedling stage, it is necessary to prevent diseases and pests such as rice blast and sheath blight. One day before transplanting rice, spray a long-lasting pesticide on the seedlings in the morning. After the seedlings are attached to the pesticide, they are planted in the field. This can effectively reduce pests and diseases in the early period of the paddy field.

3. Dense planting and transplanting in a reasonable way

Adopt a wide-narrow row transplanting method, with a width of 33.3 cm for wide rows and 23.3 cm for narrow rows, and a spacing of 20.0 cm between plants. Transplanting is done in rows, with the wide rows oriented in the east-west direction, ensuring that in the early and middle stages, light can directly reach the base of the seedlings, making full use of temperature and light. During transplantation, a sufficient number of seedlings should be planted, with approximately 170,000 seedlings per ha, ensuring a minimum of 750,000 seedlings per ha. During transplanting, pay attention to the planting method. The seedlings should be pulled out with soil and inserted immediately. Insert them straight and firmly and insert them evenly with a shallow depth of 2–3 cm. Do not insert diseased seedlings, stunted or weak seedlings, black roots, or rootless seedlings. After planting, check for missed seeds and compensate within three days.

4. Fertilize with balanced formula

(1) *Scientifically proportioned fertilization.* According to the soil fertility supply and fertilizer utilization rate, about 25 kg of pure nitrogen per ha is needed to achieve a yield of 15 t/ha, with a ratio of $N:P_2O_5:K_2O$ = 1:0.6:1.2. Rice has different nutrient absorption levels for nitrogen, phosphorus, and potassium during different growth stages, so scientific fertilization should be carried out in stages. Combine soil test results and field block seedling classification guidance to guide each mound, each farmer, and each rice growth and development stage. After the seedlings are transplanted, from the first application of tillering fertilizer to the beginning of the elongation stage, the group seedlings need to be investigated every three days. A balanced fertilizer should be added to the fields with poor growth to promote the overall balanced growth of the hundred-*mu* of seedlings.

(2) *Apply fertilizer timely and appropriately.* Fertilizers are divided into base fertilizer, tillering fertilizer, and panicle fertilizer for application.

a) Apply sufficient base fertilizer. Apply 3,000 kg of chicken manure per ha of field, 1,050 kg of 45% compound fertilizer (15-15-15), and 1,500 kg of 12% calcium-magnesium-phosphate fertilizer. During the first plowing, apply 3,000 kg of chicken manure fertilizer and 600 kg of 45% compound fertilizer at depth. For the second plowing, spread the remaining 450 kg of 45% compound fertilizer and 1,500 kg of 12% calcium-magnesium phosphate fertilizer evenly after mixing them. When plowing and harrowing in the field, attention should be paid to dry plowing, dry harrowing, and leveling with water.

b) Apply early tillering fertilizer to promote the growth of tillering. In the early stage, quick-release nitrogen fertilizer is mainly used to promote the early growth of tillering. Five to seven days after transplanting, manual weeding should be combined with the application of 112.5 kg of urea and 75 kg of 60% potassium chloride per ha. Two to fifteen days after transplanting, balanced fertilizer is applied according to different field conditions. Based on the condition of the seedlings, apply 45–75 kg of urea and 112.5 kg of potassium chloride per ha to ensure uniform growth across the entire field.

c) Reapply panicle fertilizer to promote larger panicles. This model mainly targets large panicle varieties. To fully take the advantage of easy formation of large panicles, it is crucial to apply panicle fertilizer accurately twice to promote panicle growth. Determine the application amount of panicle fertilizer based on the stem and tiller group and plant nutrition status of different fields. According to the field conditions and seedling conditions, flower-promoting fertilizer and flower-preserving fertilizer is applied in the second and fourth stages of panicle differentiation. flower-promoting fertilizer should be applied with 90–135 kg of urea, 225–375 kg of compound fertilizer (45%), and 105–135 kg of potassium chloride per ha of land. Apply 60–120 kg of urea and 112.5 kg of potassium chloride per ha for flower fertilization. To improve the grain-setting rate and grain weight, and reduce the percentage of empty husks, foliar spray Gulibao combined with disease and pest control should be applied during the heading stage and milk ripening stage, with a dosage of 30 bags per ha diluted in 750–900 kg of water. To improve grain-setting rate and grain weight, and reduce the percentage of empty husks, disease and pest control should be carried out during the heading stage and milk ripening stage. Foliar spray with grain fertilizer should be applied to the leaves, and 30 bags of fertilizer should be mixed with 750–900 kg of water and sprayed per ha of land.

(3) *Irrigate with moist air.* Throughout the entire growth period, except during the transplanting and regreening stage, the water-sensitive stage (during booting stages 5–7), panicle emergence and heading stage, and fertilizer and pesticide application, shallow water irrigation is practiced. During the rest of the period, water the field with a thin layer of water, allowing the water to naturally dry and intermittently irrigate to increase soil permeability and oxygen content, which can achieve the goal of nourishing roots with air in the early stage of seedling growth, promoting root growth and deep-rooting, improving root vitality, and delaying aging of roots. Early release of low-positioned tillering can form effective large panicles, avoiding the harm caused by long-term anaerobic conditions and toxic gases such as hydrogen sulfide and methanol produced by flooded roots, as well as avoiding excessive elongation of lower nodes due to deep water flooding, which may result in decreased resistance to lodging. In the later stage of rice growth, it is important to maintain roots and leaves, prolong the functional leaf lifespan, improve the photosynthetic capacity and fertilizer utilization efficiency of the group, and enhance grain-filling and grain plumpness. Arrange dedicated personnel to manage water inside the film and drain it in a timely manner.

The specific operation methods for water management are as follows:

a) Thin water for transplanting. During transplanting, maintain a thin layer of water to ensure the quality of transplanting and prevent shallow planting or floating. Use a row marker to mark the rows clearly before draining the field.
b) Little water for regreening. Watering with a small amount of water after transplanting for five to six days creates a relatively stable environment of temperature and humidity, promotes the growth of new roots and regening of the seedlings.
c) Shallow water and wet condition for tillering. Achieve alternating dryness and wetness, with wetness as the main condition, combined with manual weeding and irrigation of top dressing with a thin layer of water (0.5–1 cm), allowing it to dry in the air before irrigating with a thin layer of water again. Repeat this process. In clear weather, the mud water is covered, and there is no water layer in rainy weather, in order to promote root growth, promote early tillering, and reduce tillering nodes.
d) Light sun-drying for healthy seedlings germinating. When the total number of seedlings per ha reaches around 2.25 million, drainage and sun drying should be started, using multiple mild sun-drying sessions until the leaves turn yellow. If the leaf color does not turn pale when the specified period for applying panicle fertilizer is reached, continue to sun the field, and do not apply water and fertilizer. Generally, the field should be sun-dried until small cracks appear, the soil is not easily compacted, the mud surface turns white, and the leaves stand upright and fade. For fields with high groundwater levels, heavy and sticky soil texture, and vigorous seedling growth, moderate sun-drying is appropriate. Conversely, for fields with inconvenient irrigation, sandy soil, and weaker seedling growth, mild sun-drying is suitable to control the upper part, promote strong stems, increase the panicle formation rate, reduce field humidity, and mitigate pest and disease damage.
e) Ensure sufficient moisture to cultivate rice panicles. When the main stem of the rice field enters the initial stage of panicle differentiation, irrigation should be resumed, shallow water should be frequently irrigated, and the field should naturally dry out. After one to two days of drying, irrigation should be resumed in a timely manner; around the time of the panicle differentiation meiosis stage (5–7 stages of panicle differentiation), the field should maintain a water layer of about three cm.
f) Ensure sufficient water for the rice variety booting. Booting requires more water during flowering, so the field should maintain a one-inch-deep water layer to create a relatively high-humidity environment, which is beneficial for the normal growth and pollination of the booting.
g) Alternative dry-wet irrigation to make rice grains plump. From the late flowering stage to the ripening phase, a moist-dominant irrigation method should be adopted to improve root vitality and delay root aging.
h) Stop irrigation when the rice is fully mature. Five to seven days before harvest, rice enters the maturation stage. Drain the field and avoid cutting off water too early, which may affect grain-filling and yielding.

(4) *Comprehensive prevention and control of pests and diseases.* Disease and pest control should focus on prevention, and the principle of integrated prevention and control should be implemented.

The key varieties are rice thrips during the seedling stage, rice leaf rollers during the tillering stage, and rice planthopper, rice blast, and sheath blight disease, etc. Under high fertility and high population conditions, special attention should be paid to the prevention and control of sheath blight and rice planthopper in the middle and later stages. According to the pest forecast, unified prevention and control measures should be taken. Chlorantraniliprole (Kangkuan) is used to control rice thrips and stem borers, pymetrozine is used to control rice planthopper, thiamethoxam is used to control blast, and amistar is used to control sheath blight.

III. Nitrogen Reduction Cultivation Mode of Super Hybrid Rice

1. General information

There are some production problems with hybrid rice in our country, such as low fertilizer utilization, serious pest and disease damage, environmental pollution, increased production costs, and the inability of rice cultivation techniques to adapt to the full-process mechanization in the southern rice region. To solve these problems, the Hybrid Rice Physiology and Ecology Cultivation Innovation Team of the National Hybrid Rice Engineering Technology Research Center / Hunan Hybrid Rice Research Center has developed the "nitrogen-saving cultivation" technology model for hybrid rice. This technique, while reducing the total amount of nitrogen fertilizer by 20%, has resulted in a slight increase or maintenance of yield compared to the control group, with nitrogen utilization efficiency improving to over 50%.

How to reduce the use of nitrogen fertilizer while ensuring high yield and profitability? The key points of this technology include selecting good varieties, early transplanting of strong seedlings, reducing nitrogen through increasing seedlings, slow-release nitrogen fertilizer, chemical weed control, and the application of lodging resistance technology physicochemical products.

1) *Early stage.* Select good varieties. Transplant strong seedlings early. Increase seedlings and reduce nitrogen. Adapt slow-release nitrogen fertilizer.
2) *Middle stage.* Use Lingkongjie to control growth. Moderately sun-drying the field multiple times to strengthen the roots and adapt liquid silicon-potassium to promote vigorous flowering and strong stems.
3) *Late stage.* Implement comprehensive disease and pest control, adapt liquid silicon-potassium to maintain leaf health and promote strong seeds.

2. Select good varieties

Due to the diverse climates, soil, and market requirements in different rice-growing areas, the basic principles for selecting high-efficiency indica rice varieties are as follows: tillering should be abundant (>750%); earring percentage should be high (>65%); there should be many effective panicles (18,000–20,000 plants per *mu*); the leaf color should be light (nitrogen content in the top three leaves during booting stage should be 0.63%–0.70%); and the color should be good when mature.

Based on existing research, high-efficiency indica rice varieties such as Y Liangyou No. 2, Huiliangyou No. 6, and Xieyou 3026 can be selected as single-season rice varieties. Special attention should be paid to the prevention and control of rice for the listed single-season rice varieties.

3. Sow in a timely manner

In the middle and lower reaches of the Yangtze River, the sowing period for single-cropping rice in the double-cropping rice area is determined based on the ecological zone and winter cropping, generally from early April to mid-May. When choosing the sowing period for the Jianghan Plain, Dongting Lake Rice Area, and Poyang Lake Rice Area, it is advisable to arrange the booting and heading stage of single-cropping rice in mid to late August to avoid the decrease in grain-setting rate caused by high temperatures.

4. Nursery methods

For single-cropping rice cultivation, it is suitable to foster vigorous seedlings in moist seedbeds, with the seedling age controlled within 30 days.

5. Precision seeding for strong seedling cultivation

For single-cropping rice cultivation, the seeding rate is typically 0.75 to 1.0 kg/*mu*. To cultivate strong seedlings, it is necessary to sow sparsely and evenly, and the area of the seedbed should be determined according to the seedling cultivation method.

6. Transplant at the appropriate time and reduce nitrogen with more seedlings

The general principle is to increase the basic seedlings by 10% compared with normal cultivation and reduce nitrogen fertilizer by 20%, that is, "increase seedlings by 10% and reduce nitrogen by 20%."

Sow single-cropping rice plants two seedlings per hole. Guarantee 11,000 to 15,000 plants per *mu*.

7. Apply reduced nitrogen fertilizer with a combination of quick- and slow-release effects

The total amount of nitrogen fertilizer should be determined based on different yield targets.

For single-cropping rice with a yield of 900 kg/*mu*, apply nitrogen (equivalent to pure nitrogen) at a rate of 15.2 to 16.0 kg/*mu*, with an N:P:K ratio of approximately 1:0.6:(1–1.2).

Base fertilizer adopts the application of disposable slow-release compound fertilizer (combining quick-release and slow-release effects), with a total nitrogen reduction of 15% to 20% compared to conventional fertilizers. The nitrogen fertilizer dosage for different seasons is shown in TABLE 7.1. For early-season rice, the ratio of base fertilizer to topdressing is applied at 6:4, while for late-season hybrid rice and single-cropping rice, it is applied at 4:6. Since single-cropping rice is mostly a variety with large panicles and fewer seedlings in the early stage, the proportion of grain fertilizer application should be increased to reduce nitrogen loss in the field.

TABLE 7.1 Recommended reduced-nitrogen fertilizer application for super hybrid rice

Season	Recommended reduced nitrogen fertilizer (kg/*mu*)			Regular fertilizer (kg/*mu*)		
	Pure nitrogen	P_2O_5	K_2O	Pure nitrogen	P_2O_5	K_2O
Single-cropping Rice	15.2–16.0	7.5–8.0	16.0–18.0	18.0–20.0	7.5–9.0	18.0–20.0

Note: Amount in this table is not suitable for varieties with high fertilizer tolerance.

8. Manage water scientifically and adjust fertilizer with water

Transplantation should be followed by consistent shallow watering until the early tillering stage, maintaining a water depth of two to three cm in the field. This practice aims to use water for nutrient regulation and to promote nutrient uptake by rice. During the middle stage, irrigation should be combined with sprinkling, with shallow-water irrigation as the main method. When the number of effective tillers per *mu* reaches 90% of the total tillers, the field can be drained and exposed to the sun. During the later stage, water should be provided for booting. The seeds should be kept moist, with intermittent watering. Stop watering five to seven days before maturity, but do not stop watering too early.

9. Promote inhibition, prevent lodging and declining

For single-cropping rice varieties with high plant height, spray 40 g of Lifengling and 200 mm of liquid silicon potassium per *mu*, diluted with 35–40 kg of water, evenly sprayed five to seven days before the elongation stage (can be combined with pest control). This not only increases the lodging resistance of rice plants in the later stage but also reduces plant height and decreases the number of grains in an appropriate way, achieving high yields with lodging resistance.

10. Comprehensive control of diseases and pests

The key is to control "three diseases and three pests" including sheath blight, rice blast, and false smut; stem borer (Chilo suppressalis and Tryporyza incertulas), leaf roller, and rice planthopper. Specifically, the application of targeted pesticides should be timely carried out according to the requirements of local disease and pest monitoring and reporting departments to prevent and control diseases and pests.

IV. Cultivation Mode of Super Hybrid Rice Lodging Resistance

1. General information

Lodging is one of the main factors causing reduced yield and quality decline in rice, and it also significantly increases harvesting costs. Lodging mainly consists of two types: lodging in the roots and lodging in the stems. The occurrence of lodging mainly happens during the grain-filling stage and later until the muturing phase. In China, lodging causes a yield reduction of 10% to 30% annually. To address the above issues, the Hunan Hybrid Rice Research Center has developed a hybrid rice system called lodging resistance cultivation techniques with "combine inhibition and promotion, strengthen stems for higher lodging resistance" as its core. By applying this technological achievement, the stem lodging resistance can be increased by 30% to 40%, and the yield can be increased by more than 5%.

2. Reasons for lodging of rice

1) *The variety itself has weak lodging resistance.* For example, the plants are too tall, the length between the first and second internodes at the base is too long (more than five cm for the first internode), the plant type is not compact, and the stem is thin and not strong (such as some high-quality indica rice).
2) *Improper fertilization practice.* Improper fertilization such as excessive nitrogen application at the expense of phosphorus and potassium and lack of attention to balanced fertilization, can lead to imbalanced growth. The use of a "one-size-fits-all" fertilization approach results in vigorous early growth, over-elongation, and thinness of stem internodes at the base and weakened stems later in the season, leading to premature senescence.
3) *Improper water management.* The deep-water layer and loose soil in the field lead to poor rice root growth or shallow root distribution. Untimely and ineffective field-drying, early drainage, and dehydration lead to premature senescence of rice plants in the ripening phase.
4) *Improper pest and disease control.* When there are poor control effects of sheath blight and rice planthopper, sheath blight can cause the basal leaves to lose their root function severely, and rice planthopper can directly lead to stem withering when severe.
5) *Inadequate coordination between field management and planting methods.* Generally, direct-seeding rice and transplanting cultivation often result in poor root development, shallow and unstable rooting, poor soil fixation ability, and susceptibility to flat lodging due to wind and rain invasion. In the case of poor land preparation quality and short flooding time, machine-transplanted seedlings are prone to deep planting, poor stem base filling, thin cell walls, and low cellulose content, resulting in lodging.

3. Principle of rice lodging resistance technology

This technique is based on understanding the morphological and mechanical mechanisms of lodging resistance in super hybrid rice under conditions of super-high yield. It not only improves the photosynthetic capacity of the group and delays leaf senescence to promote healthy roots and prevent lodging but also increases the storage of substances in the stem sheath and enhances the ability of stem elongation, achieving high yields with lodging resistance. Regardless of whether the plant height is tall or short, the short and strong basal internodes of rice, reasonable arrangement of elongated internodes, and high silicon, potassium, and cellulose content contribute to significantly improving the rice's resistance to lodging. Therefore, any variety that meets the requirements of short basal internodes and reasonable arrangement of elongated internodes, as well as technical measures that promote healthy roots and strong stems and regulate the length of basal internodes and the arrangement of elongated internodes, can achieve high yield with lodging resistance.

4. Region suitable for technology

The technology is applicable to various types of rice production areas in the south.

5. Specific technical measures

1) *Adapt to local conditions and choose good varieties.* Field morphological observation has shown that varieties with good plant type, strong stems, and short internodes at the stem base generally have strong resistance to lodging.
2) *Sow in a timely manner and cultivate strong seedlings.* Choose the appropriate sowing time and method. Sow sparsely and evenly and cultivate multiple tillers and strong seedlings.
3) *Scientific water management brings about strong roots and sturdy stems.* Scientific water management and good root growth are the keys to preventing lodging. Generally, in the early stage of transplanting to tillering, shallow water is frequently irrigated; then, irrigation and dewing are combined during the medium stage, and shallow irrigation is the main factor. When the number of seedlings per *mu* reaches 80% of the expected effective panicles, lower the water level and sun-dry the field. Ensure there is sufficient water for booting and promote grain-filling through alternative dry-wet irrigation in the later period. The field should be drained five to seven days before maturity. Water cannot be cut off too early.
4) *Adapt chemical regulation with a combination of inhabitation and promotion.* For varieties with high plant height that are easy-lodging and high-quality rice varieties with fine and soft stems, it is appropriate to adopt the regulation method of shortening the basal internode by inhibition, spraying 35–40 g of Lifengling with 30 kg of water five to seven days before stem elongation stage, and spraying evenly. Do not spray repeatedly during spraying.

   For varieties with shorter heights but weaker lodging resistance capability, only use the method of promoting regulation to improve lodging resistance force. Directly spray 200–300 ml liquid silicon-potassium fertilizer one to two times per *mu* on the leaf surface during the internode elongation period (it can be mixed with pesticide spraying). At the same time, liquid silicon-potassium fertilizer can significantly improve heat resistance, with grain-setting rate generally increasing by 3% to 5%, and the effect is particularly prominent when encountering high-temperature hazards.
5) *Adopt a comprehensive, cost-saving, and effective method to control diseases and pests.* Use corresponding pesticides in a timely manner, according to the local disease and pest situation report. Pay special attention to the prevention and control of sheath blight and rice planthopper.

CHAPTER 8

Evaluation of Climatic Ecological Adaptability of Super Hybrid Rice

The evolution and turnover of super hybrid rice varieties from high yield to super high yield is, in essence, the adaptation of super hybrid rice to complete its life cycle of growth and development until maturity by using the climatic ecological resources conditions such as light, temperature, water, fertilizer and gas given by the ecological environment.

Therefore, two main factors primarily affecting the productivity of super hybrid rice are the ability of super hybrid rice to utilize ecological resources and transform them into yield, which is mainly regulated by the genotype and genetic background of varieties, and the ability limit of provided ecological and climatic resources which can be utilized within a certain time and space. The resources include ecological resources both given by nature and man-made resources. Social activities regulate ecological resources, mainly the influence of human cultivation technology measures. Therefore, the rational use of local agricultural resources and the selection of cultivation systems and super hybrid rice varieties suitable for local climatic conditions are important ways to scientifically plant fields and improve the yield of super hybrid rice. The productivity of super hybrid rice varieties not only depends on the genetic characteristics of the varieties but also is closely related to the ecological resources in time and space during the whole reproductive period and the adaptability of varieties to ecological resources. Therefore, the regionalized planting of super hybrid rice varieties should give priority to those suitable for local ecological and climatic conditions.

The climate varies significantly from places in China to its vast territory. Given such, various planting systems for super hybrid rice have been formed, mainly including single-cropping, double-cropping, triple-cropping in one year, and triple-cropping in two years. Single-cropping rice is mainly cultivated in the northern regions due to low temperatures, limited precipitation, and relatively short growing seasons. In the southern regions, the temperature is higher, the rainfall more abundant, and the growing seasons longer. The double-cropping rice is distributed broader, primarily in the south of N 90° parallel. Super hybrid rice in the Yangtze River Basin and the southern China region is often intercropped with other crops, forming various crop rotation systems.

Ecological adaptability is the ability of a species to achieve equilibrium with dynamic environments. The ecological adaptability of super hybrid rice emphasizes its ability to adapt to different ecological conditions and resilience to adverse conditions while maintaining stable and high yields. The ecological adaptability of modern rice includes adaptability to geographical location, temperature, water availability, light intensity, and soil fertility.

I. Characteristics of Super Hybrid Rice Adapted to Latitude

In China, rice cultivation areas span up to 40° latitudes. Indica subspecies rice is mainly distributed in lower latitude areas, and japonica subspecies rice is in areas with high latitude and altitude. Super hybrid rice in Yunnan Province shows obvious altitudinal zonation. Indica rice is mainly planted in areas at altitudes of 76 to 1,400 m, and japonica rice is mainly planted at altitudes of 1,600 to 2,700 m. Both indica and japonica rice can be planted in the area with altitudes from 1,400 to 1,600 m. It indicates that the differentiation of indica and japonica has a close relationship with both latitude and altitude. As latitude increases from south to north, the upper limit of super hybrid rice cultivation decreases; while as longitude increases from east to west, the upper limit of cultivation increases.

It is reported that the "Super Hybrid Rice 'Hundred, Thousand, and Ten Thousand' High-Yielding Tackling Demonstration" guided by Yuan Longping, an Academician of the Chinese Academy of Engineering, has set up key research bases with hundreds of *mu* each in more than 80 counties and cities nationwide. In 2016, an expert group measured the yield of the rice in the demonstration base located in Dadian Town, Junan Town, Linyi City, Shandong Province (E 118.77°, N 35.5°). The super hybrid rice grew evenly, had large panicles and many grains with a high grain-setting rate, and no major diseases. It was measured that the average yield was 15.21 t/ha. In 2017 the average yield reached 15.41 t/ha, which broke the highest yield record of rice in the northern high latitudes (about N 36°). At the same time, the super hybrid rice base at Rizhao, Shandong, in 2016 had an average yield of 14.71 t/ha. This indicated that super hybrid rice had strong growth adaptability at around N 36° and that ultra-high yield targets can also be achieved by strengthening cultivation management.

Super hybrid rice exhibits strong ecological regional characteristics. This indicates that super hybrid rice can only unleash its potential for ultra-high yield production under extremely suitable geographical, environmental, climatic, and soil conditions. Therefore, under general ecological conditions, a significant increase in yield is hardly achieved, even with increasing production inputs. Thus, from the perspective of cultivating ultra-high yields of super hybrid rice, setting production targets is not a case of the more, the better. Instead, local ecological and geographical conditions should be fully considered for adaptation to local conditions being crucial. The factors constituting the yield of super hybrid rice exhibit diversity, and obtaining large-scale ultra-high yields requires the harmonization of various yield-forming factors.

For example, super hybrid rice should be sowed earlier in high latitude areas (thin film or greenhouse seedling methods can be used to avoid cold), which can ensure that the later growth stages, especially reproductive growth, occur in warm seasons, thus guaranteeing yield. In addition, high-yielding and high-quality varieties should be selected, which have medium panicle, higher thousand-grain weight, smaller leaf angle, compact plant type, strong tillering ability, strong disease and cold resistance, and high grain-setting rate.

II. Characteristics of Super Hybrid Rice Adapted to Altitude

1. The planting altitude range of super hybrid rice

The upper limit of rice cultivation altitude in the world is in the subtropics rather than the tropics. Therefore, the upper limit of rice cultivation in the subtropics is generally higher than in the tropics. In the high plateau river valley basins of the western subtropics of China, the upper limit of planting super hybrid rice reaches 2,600 m to 2,700 m. In small areas of Ninglang County, Yunnan Province, and Yanyuan County, Sichuan Province, the altitude ranges from 2,630 to 2,710 m, making it the highest altitude for rice cultivation in the world. Different types of super rice varieties have different temperature requirements. Due to differences in geographical environments and temperature gradients, the upper limit of planting altitude for super hybrid rice also varies. In the southeastern part of China, the upper limit for cultivating indica rice is 1,100 m, while it rises to 2,100 m in the southwest. For japonica rice, the upper limit in the southeast is 1,200 m, while it reaches 2,710 m in the southwest. The planting altitude limit of hybrid rice is closely related to the parental lines, indicating that the planting altitude limit of both indica hybrid rice and japonica hybrid rice is close to or lower than that of ordinary rice and japonica rice.

2. The Impact of altitude on yield and its components in super hybrid rice

Altitude is an important factor influencing the yield of super rice. The ecological environment of different altitudes varies significantly, which in turn affects the growth and development of super hybrid rice, leading to considerable differences in yield. In the Mianyang area of Sichuan Province, within an elevation range of 400 to 1,400 m, the yield of super hybrid rice, as well as the number of productive ears and grain number per ear, will first increase and then decrease with rising altitude. The same variety of rice planted in Yunnan's Taoyuan (elevation 1,170 m) has 50% to 60% more panicles than in Fujian's Longhai (elevation 652 m), resulting in higher yields. In Xichang, Sichuan (elevation 1,580 m), the high-yielding fields of super hybrid rice show a greater number of productive ears compared to the basin rice areas, indicating that a higher number of productive ears is a crucial reason for the higher yields in mid- to high-altitude regions such as Yunnan's Taoyuan and Binchuan, and Sichuan's Xichang.

Research on the accumulation and distribution of above-ground dry matter in super hybrid rice at three different altitudes in Yunnan (400 m, 1,900 m, and 2,400 m) found that the total dry matter production during the entire growth period was highest at 1,900 m above sea level, followed by 400 m, and lowest at 2,400 m. Both the ratio of panicle weight to total weight and the ratio of panicle weight gain to total weight gain were higher in low-altitude areas compared to high-altitude areas. The reason for decreasing grain-setting rate with increasing altitude may be the decrease in temperature affecting the transport of assimilates to the grains.

Research on the growth effects of different types of super hybrid rice varieties at five different altitudes (1,000 m, 1,180 m, 1,260 m, 1,410 m, and 1,490 m) in Leishan County, Guizhou Provience, found that all the test varieties could grow and develop normally, except those in 1,490 m altitude. However, with increasing altitude, temperatures will decrease, leading to longer growing periods and lower grain-setting rates.

In Longhui County, Hunan, three super hybrid rice combinations (Y Liangyou 900, Xiangliangyou 2, and Shenliangyou 1813) were planted at four different altitudes (300 m, 450 m, 600 m, and 750 m).

The average yields at different altitudes show the following trend: 450 m (11.45 t/ha) > 300 m (9.93 t/ha) > 600 m (9.63 t/ha) > 750 m (7.77 t/ha), indicating yields decreased significantly with increasing altitude from 600 m. Considering the yield of individual combinations at the four altitudes, Y Liangyou 900 (10.36 t/ha) > Xiangliangyou 2 (9.65 t/ha) > Shenliangyou 1813 (9.08 t/ha). Y Liangyou 900 showed the best adaptability and highest yield, followed by Xiangliangyou 2. Additionally, starting from 600 m altitude, plant height, panicle length, total number of grains, and grain-setting rate all significantly decreased with increasing altitude.

The study shows that as altitude increases, the varieties present a longer pre-heading stage, post-heading stage, and whole growth stage. The impact of climate on yield at different stages of growth is generally similar among different varieties, and the overall trend of climate impact on yield structure for different types of rice varieties is also similar. The high-yielding variety Y Liangyou 900 is well-matched with the climate at altitudes of 300 to 450 m in Longhui, facilitating the realization of its yield potential. The cold-resistant variety Xiangliangyou 2 matches the climate at altitudes of

300 to 600 m in Longhui, while Shenliangyou 1813 matches the climate at altitudes of 450 to 600 m in Longhui, allowing them to make full use of thermal resources and achieve higher yields.

The magnitude of the impact of altitude on different super hybrid rice varieties was as follows: indica rice > indica hybrid rice > japonica rice. According to the degree of impact of altitude on different rice varieties, it is recommended to plant indica rice in areas below 1,100 m altitude, indica and japonica rice in areas at 1,100–1,300 m altitude, japonica rice in areas at 1,300–1,450 m altitude., As for areas above 1,450 m, either no rice or extremely cold-resistant varieties can be planted.

Currently, the super hybrid rice growing areas in China are mainly distributed in the Yangtze River Basin, South China, and the Huang–Huai Basin whose altitude is generally below 1,100 m. Research on high-yielding super hybrid rice in low-altitude areas of the Dabie Mountains in Anhui (63 m) indicates that super hybrid rice exhibits high-yielding characteristics in low-altitude regions. The number of reproductive panicles reaches 2.52 to 2.62 million per ha, with a total grain number per panicle ranging from 220 to 254, a grain-setting rate of 83% to 86%, a thousand-grain weight of 28 g, and an average yield of 12.02 t/ha.

In a study on the cultivation of super hybrid rice Luliangyou 106 in low-altitude (618 m) and high-altitude (1,037 m) regions of Guizhou, it was found that as altitude increases, the daily average temperature decreases, the accumulated temperature and sunlight required for the growth of super hybrid rice increase, the tillering rate slows down, the growth period significantly extends, plant height decreases significantly, panicle length noticeably shortens, total grain number significantly decreases, reproductive ear number slightly increases, and the grain-setting rate and thousand-grain weight decrease to some extent. The combined effects of these yield factors lead to a significant decrease in the yield of super hybrid rice in high-altitude areas.

Therefore, within a certain altitude range in a place, the yield of super hybrid rice decreases with increasing altitude. For example, in Sichuan, within the range of 400 to 1,400 m above sea level, the yield of super hybrid rice shows a trend of first increasing and then decreasing with increasing altitude. In Guizhou, with increasing altitude, hybrid rice shows a lower yield within 618 m to 1,307 m altitude, while it shows a longer growth period and lower both grain-setting rate and yield within 1,000 m to 1,410 m altitudes. In Jiangxi (230 to 830 m altitudes), the yield of super hybrid rice decreases as altitude increases. In order to mitigate the impact of altitude on rice cultivation, corresponding agricultural measures should be taken, and varieties with strong adaptability should be selected. The auxiliary cultivation of super hybrid rice can increase soil temperature to enhance low-temperature resistance, thereby achieving high yields.

In conclusion, high yield can be achieved when super hybrid rice is planted in areas at suitable altitude.

III. Differences in High-Temperature Resistance and Adaptability of Super Hybrid Rice

1. Characteristics of temperature changes

From a global perspective, over the past 100 years, climate anomalies have led to frequent changes in climate factors such as temperature, precipitation, and rainfall, which have had a great impact on super hybrid rice production. According to the 2012 assessment report of the Intergovernmental Panel on Climate Change (IPCC), global land and ocean surface temperatures have risen by 0.85°C from 1880 to 2012, with the last 30 years (1983–2012) being the warmest 30 years with the highest temperature rise in the northern hemisphere.

In China, a study on temperature changes at 745 stations over the past 50 years (1963–2012) showed that, with the exception of the slight cooling trend in the southwest region, all other regions of the country showed a clear warming trend, with the northern regions experiencing more pronounced warming. Since 1978, the temperature in China has shown a gradual upward trend, with a more significant increase in the northern regions and more warming in winter than in summer. China's surface temperature has increased by 0.2°C to 0.8°C in the past century, which is greater than the global average. It has increased by 0.6°C to 1.1°C in the past 50 years, and it is predicted that the surface temperature will increase significantly in the next 20 to 100 years. By analyzing the variation characteristics of the daily temperature range in China from 1952 to 2001, it is found that the daily temperature range shows a decreasing trend, with significant differences in different regions and seasons, especially in the middle and high latitudes. Climate studies in Shaanxi Province over the past 54 years (1961–2014) show a significant increase in average annual temperature in northern Shaanxi, Guanzhong, and southern Shaanxi regions, with an increase of 0.21°C every 10 years. The annual accumulated temperature above 10°C has also increased significantly over the past 50 years, reaching 59.2°C every 10 years. The average annual temperature in Hubei Province showed an overall upward trend over the past 51 years (1959–2009), with a climate tendency rate of 0.195°C per decade, showing a decreasing trend from southeast to northwest. The average annual temperature and the average minimum temperature in Zhejiang Province have both shown a significant upward trend over the past 50 years (1961–2010), with the effective accumulated temperature during the growing season increasing significantly by 3.04°C per decade.

As a result, climate warming has had a great impact on the production process of super hybrid rice, causing changes in the hydrothermal conditions of the super hybrid rice region and the layout and planting structure of super hybrid rice, which has also had a profound impact on all aspects of the growth and development of super hybrid rice.

2. Characteristics of changes in rainfall

According to the IPCC 2007 report, rainfall is greatly influenced by latitude factors. Rainfall performance is increasing in middle and high-latitude regions, while lower-latitude regions are showing a declining trend. Some studies have analyzed the pattern of annual and seasonal rainfall changes in the northern hemisphere over the past 130 years (1855–1984), indicating that the region of N 5° to 35° showed a downward trend, while N 35° to 70° showed an increasing trend, and precipitation in the northern hemisphere changed greatly from year to year. Since the twentieth century, total rainfall on the Earth's land surface has increased by an average of 1%, and there is an increasing trend; the total rainfall in the N 30° to 85° of the northern hemisphere has increased by 7% to 12%, a remarkable increase, and the region of S 5° to 50° in the southern hemisphere has increased by 2% to 3%. However, in the mid-1980s and 90s, rainfall in the northern hemisphere at N 30° showed a clear decline. Rainfall has increased markedly in most parts of the world, and the number of extreme value drops has increased markedly.

The rainfall in different regions in China generally showed an increasing trend. Analysis of daily precipitation data from 739 meteorological stations in China from 1951–1999 shows that the frequency of heavy rainfall in northwest China was an increasing trend, while northern China showed a declining trend. A study on rainfall trends in western and eastern China from 1951 to 2003 found regional differences in rainfall in western and eastern China, mainly the eastern region being greatly affected by atmospheric circulation. Studies have shown that over the past 50 years in my country, areas with high rainfall have shifted from North China to South China, first to the middle and lower reaches of the Yangtze River, and then south to South China. The analysis of the characteristics of rainfall changes in southern China in the past 60 years (1956–2015) shows that the annual rainfall in Hubei, Zhejiang, Jiangxi, Hunan, Guangxi, and other regions reaches 1,144.59–1,584.43 mm, indicating that the annual rainfall in these regions is relatively large, and the regional Water resources are abundant. In the past 50 years (1961–2010), the average annual rainfall in Zhejiang Province showed an increasing trend. The proportion of precipitation above heavy rain and the interannual changes in average daily precipitation intensity showed a clear upward trend.

3. Characteristics of sunshine changes

The number of sunshine hours in most regions of the country has been decreasing to different degrees. Numerous studies have indicated a significant decrease in sunshine hours across the country. For example, the average annual sunshine hours in Henan Province over the past 53 years (1960–2012) has shown a clear trend of decrease, with a reduction of 8.5 hours per decade. Over the past 59 years (1951–2009), the sunshine hours nationwide have shown a significant decreasing trend, with an average reduction rate of 36.9 hours per decade. In the past 47 years (1961–2007), the number of sunshine hours in most places in Northwest China has decreased significantly, with an average decrease rate of 19.92 hours per decade. The average annual sunshine hours in Yunnan Province in the past 44 years (1969–2012) have generally shown a fluctuating downward trend, with a decline of 14.95 hours per decade. The average annual sunshine hours in Shandong Province in the past 40 years (1970–2009) have shown a significant decreasing trend, with a typical distribution feature of less in the south and more in the north. Climate change studies on Changsha Station in Hunan for the past 28 years (1987–2014), Jishou Station for the past 23 years (1992–2014), and Changning Station for the past 22 years (1993–2014) show that the average sunshine hours in Hunan are distributed with less in the west and more in the east. The areas with high values of sunshine hours include the Dongting Lake area, the Changsha–Zhuzhou–Xiangtan area, the central-eastern part of Loudi, and the northeastern part of Shaoyang.

4. Impacts of climate change on super hybrid rice production

In the context of climate warming, studying the characteristics and development trends of agricultural climate change and conducting an in-depth analysis of the impact of climate change on the growth of super hybrid rice is of great practical significance to the layout of super hybrid rice in various regions and the adjustment of planting systems and cultivation measures.

Climate change directly alters regional hydrothermal changes and affects the environment in which super hybrid rice grows. To a certain extent, the growth of super hybrid rice is a cumulative effect of temperature and moisture. Temperature changes will also indirectly affect the production of super hybrid rice by affecting the length of the growing season. Due to the regional nature of climate change itself, as well as the regional characteristics of super hybrid rice cultivation in China, the impact of climate change on super hybrid rice production is not entirely consistent. Locally speaking, higher temperatures will increase the opportunities for super hybrid rice to utilize light and heat conditions, while sufficient precipitation also provides a foundation for the growth of super hybrid rice (TABLE 8.1).

TABLE 8.1 Impact of climate change on super hybrid rice production in different ecological regions

Ecological regions	Impacts
North China	Warmer and wetter conditions are favorable for the production of super hybrid rice, while warming with dryness is unfavorable, and precipitation may decrease fields.
Southwest China	Some areas have experienced cooling in recent years, which is unfavorable for super hybrid rice, but future warming climate is advantageous for super hybrid rice production.
Central and Eastern China	Warming increases the risk of damage during the flowering and grain-filling stages of rice, resulting in reduced yield and quality. It is favorable for improving planting conditions and super hybrid rice production, while wetter climate is unfavorable for production.
South China	There is relatively less warming with a trend toward increased humidity. This increases the risk of heat stress in southern rice-growing areas. Early rice yields decrease more in the southwest, while late rice yields decrease more in the northwest, with less decrease in the south.

Climate change plays a crucial role in rice production and even food security. As a major rice producer and consumer, the impact of climate change on China's rice production will be directly related to the food security of the world's population.

The core impact of climate change on rice production is its impact on yield. Regardless of changes in temperature, precipitation, or the associated rice planting environment, rice production is the most important focus of attention. Under assumptions of future climate scenarios, changes in rice yield in China show significant regional variations. The impact of climate change on rice yield is highly uncertain, as it may lead to yield increases due to factors such as temperature increase during the growing season and fertilization effects, etc., or yield decreases due to reasons such as extreme temperatures.

From a long-term perspective, super hybrid rice does not always passively accept all the changes brought about by climate change, but due to its adaptability, it gradually assimilates with the environment, mitigates the negative effects of climate change, and adapts to the new climate environment. Measurements, such as the replacement of super hybrid rice varieties, changes in sowing time, etc., are implemented to adapt to high temperatures or longer growing seasons, fully utilizing light and temperature resources, and going after profit while avoiding harm.

In the context of climate warming, the length of the growing season for super hybrid rice in China has changed significantly, mainly manifested as an earlier suitable sowing time or a delayed safe maturity time. In the 40 years from 1961 to 2000 in China, the climate growing period increased by an average of 6.6 days nationwide, by an average of 10.2 days in the northern region, and by 4.2 days in the southern region. Over the past 50 years, the northern boundary of double-cropping rice cultivation across the country has continuously moved northward, making areas originally cultivated with single-cropping rice or rice-wheat rotation suitable for double-cropping rice cultivation, expanding the scope of double-cropping rice cultivation and theoretically increasing the yield of rice.

For every 1°C increase in the average temperature during the growth period of super hybrid rice, the actual length of the growth period shortens by nearly eight days. With the temperature expected to continue rising in the future, it will further shorten the growth period of super hybrid rice, causing the actual growth period line in the eastern and northeastern regions to move northward.

5. Evaluation of heat tolerance in rice and the impact of high-temperature stress on super hybrid rice

(1) *Evaluation of high-temperature stress in super hybrid rice.* High-temperature stress refers to the harm caused by extremely high temperatures to the growth, development, and yielding of super hybrid rice. The indicators of high-temperature stress vary due to different research purposes and methodologies (TABLE 8.2).

TABLE 8.2 The indicators of high-temperature stress on super hybrid rice

Research area	Indicator	Source
Indoor experiment	32°C is the damaging temperature for thousand-grain weight and 35°C for grain-filling rate.	1976, Shanghai Institute of Plant Physiology
Yangtze River Basin	Daily average temperature over 30°C for 3–5 consecutive days, or temperature exceeding 35°C for more than three days (5 hours/day)	2007, Tian Xiaohai, etc.
Nanjing, Anqin	≥39°C for 5 hours nonstop, or ≥35°C for three consecutive days (5 hours/day)	2007, Zheng Jianchu, etc.
Hubei Province	The highest ≥35°C for consecutive days	2009, Zhang Fangfang, etc.

Extreme temperatures can be divided into extreme low temperatures and extreme high temperatures, among which the latter has a destructive effect on the growth and development of super hybrid rice, posing a threat to the high and stable yield. The booting and heading stages of hybrid rice are critical periods for yielding. The two stages when hybrid rice is most vulnerable to extremely high temperatures meet in summer in China. The increasing frequency and intensity of extremely high-temperature events have a very serious impact on the yield of hybrid rice.

(2) *Research on the high-temperature tolerance evaluation of super hybrid rice.* The study suggests that differences in heat tolerance of hybrid rice can be identified through a combination of artificial climate chambers and field natural treatments. Among them, the field natural evaluation method refers to the method of the International Rice Research Institute, which uses natural temperature treatments under field conditions to evaluate the heat tolerance of different hybrid rice varieties. Heat tolerance of rice can be identified when the daily average temperature reaches 30°C or when the daily maximum temperature is achieved. By using the grain-setting rate as the evaluation index for heat resistance/tolerance of varieties, heat-resistant/tolerant varieties are classified into three categories: good, moderate, and poor. The temperature threshold difference between heat-resistant/tolerant varieties and heat-sensitive varieties is generally 3°C to 4°C.

(3) *The effect of high temperature on yield.* As China's climate is continuously warming, frequent occurrences of extreme high temperatures significantly impact the production of super hybrid rice. The heading and grain-filling stages are critical for the development of hybrid rice and play a decisive role in its yield formation. These stages are also the most vulnerable period for extremely high temperatures during its life cycle. A major characteristic of climate warming is the fluctuation in extreme temperatures. With the increased frequency and intensity of extreme temperature events under climate change, the overlap between high-temperature occurrences and vulnerable stages of rice will directly reduce the yield. Various regions in China have experienced severe damage to super hybrid rice production. For instance, from 1983 to 2010, there was a significant increase in the frequency of high-temperature damage in the northeastern part of Jiangxi, northern areas of Ganzhou, the Jitai Basin, and the Ganfu Plain. In Jiangsu, from 1998 to 2009, the significantly rising average tempera-

ture caused prevalent high-temperature damage in the southern part of the province. In Anhui and Zhejiang, since the mid-1980s, there has been a noticeable increase in high-temperature damage to early-maturing rice in areas along the Yangtze River, Jianghuai region, and the Jinhua–Lishui area. In the eastern and central parts of the Sichuan Basin, high-temperature damage on rice has shown a worsening trend over the past 20 years, particularly in the northeastern part of Chongqing where severe heat damage has increased significantly. Cases in which super hybrid rice production has been seriously harmed have occurred from time to time in various parts of China. For example, from 1983 to 2010, the frequency of heat damage in northeastern Jiangxi, northern Ganzhou, the Jitai Basin, and the Ganfu Plain increased markedly. The average temperature in Jiangsu increased significantly from 1998 to 2009, mainly due to high temperatures and heat damage in southern Jiangsu. After the mid-1980s in the Anhui and Zhejiang regions, there was a marked increase in heat damage to early-season rice along the river in Anhui, the Jianghuai region, and the Jinhua–Lishui area of Zhejiang. The high-temperature heat damage of rice in the eastern and central Sichuan Basin has been increasing in the past 20 years, and the severe heat damage in the northeast of Chongqing has increased significantly. The frequency and intensity of heat damage increased during high-temperature vulnerable periods for rice in most of northeastern China, most of the middle and lower reaches of the Yangtze River, and western parts of southern China. In parts of the middle and lower reaches of the Yangtze River and parts of northern South China, the temperature was above 38°C (lasting about 20 days), which is rare, from June to August 2003, and more than 466,000 ha of single-cropping rice in Hubei were affected; the entire Yangtze River Basin is conservatively estimated to have lost 52 million tons of rice. In 2013, Yiyang, Hunan Province also experienced hot weather for more than 40 days (the highest was 41.9°C), causing early-maturing rice to suffer from "heat-forced maturity" and seriously cut production.

The main impact of high-temperature heat stress is a reduction in the grain-setting rate of hybrid rice. Studies have shown that exposure to heat stress during the vegetative and reproductive growth stages of hybrid rice can lead to damage to potential repositories such as panicle number and spikelet number, resulting in a decrease in grain-setting rate. Hybrid indica rice subjected to 35°C stress treatment exhibits decreased pollen viability, pollen germination rate, and grain-setting rate, with the decline becoming more severe with increasing stress temperature and duration. Grain-setting rates during the heading stage significantly decrease after exposure to high-temperature heat stress ranging from 35°C to 41°C, with the degree of decrease in setting rate increasing with higher stress temperatures and longer durations. Rice subjected to high-temperature heat stress shows an average decrease in grain-setting rate per panicle of nearly 7% compared to the control, with a decrease in thousand-grain weight of 2.1 g. High temperatures affect rice grain development and alter rice quality, leading to varying degrees of decline in milling quality, appearance quality, eating quality, and nutritional quality.

During the booting stage, high temperatures can significantly reduce the above-ground total dry weight, panicle dry weight, and their distribution indices of hybrid rice, while high temperatures during meiosis will lead to a decrease in spikelet number per panicle, grain-setting rate, grain weight, and yield. With increasing temperature and duration of high-temperature treatment during the booting stage, the photosynthetic rate and yield decrease more significantly. After high-temperature stress, there is a decreasing trend in spikelet number, grain-setting rate, and thousand-grain weight, with the scales of impact being grain-setting rate > spikelet number > thousand-grain weight. High temperatures during the heading and flowering stage significantly reduce the grain-setting rate of hybrid rice, but the impact on thousand-grain weight and per panicle grain number is relatively small. High temperatures during different growth stages significantly reduce grain yield, with the highest average reduction of 91.3% during the entire growth period, and the heading stage being the most sensitive to high-temperature stress, leading to a reduction of 58.5%. High temperatures during the late growth stage of hybrid rice accelerate the early grain-filling rate, shorten the grain-filling time, affect the grain-filling process, and consequently reduce thousand-grain weight. High-temperature stress during the heading and grain-filling stage significantly reduces the grain-setting rate and grain weight of hybrid rice, resulting in a severe decrease in yield.

The regional climate ecology of super hybrid rice in China has caused great uncertainty in the yield of super hybrid rice planted in different regions in different years. At the same time, the extremely high temperature may also cause weeds, insect pests, and diseases around the rice fields, indirectly affecting the yield of super hybrid rice. From the perspective of species evolution, the evolution of super hybrid rice species is that they actively adapt to ecological changes, gradually assimilate with the ecological environment, and mitigate the negative effects of the ecological environment, thereby adapting to new climate and ecological changes, rather than passively accepting negative changes. For example, from the first phase to the fifth phase, super hybrid rice varieties changed, and the yield increase from 700 to 900 kg is the result of the excellent variety of resources of super hybrid rice adapted to the changing ecological environment.

(4) *The effects of high temperature on the growth and development of hybrid rice.* High-temperature stress occurs during different growth stages of hybrid rice, with varying impacts on its growth. Different combinations show significant differences in their ability to tolerate high temperatures. The period with the greatest impact on yield is the flowering stage, followed by panicle differentiation, booting, and early grain-filling stages. Studies have shown that the most vulnerable periods of hybrid rice to high temperatures regarding fruit set are meiosis and flowering. High temperatures during the booting stage can cause abnormal pollen development, hinder the development of branches and spikelets, reduce the number of spikelets, and delay booting. High temperatures during the booting and heading stages hinder pollen development, leading to poor fertilization. High temperatures during the microspore formation stage will cause abnormal development of pollen mother cells, and delay tetrad dispersal. At the pollen grain maturation stage, abnormal morphology of the anther wall and pollen grains leads to abortion. High temperatures during the heading stage reduce the ability of pollen grains to expand, leading to obstruction of anther dehiscence and decreased pollen and stigma vitality, resulting in reduced grain-setting rates. High temperatures before flowering cause severe degradation of starch inside pollen grains, reducing pollen vitality, while high temperatures during flowering cause pollen grains to shrink and decrease in diameter. From the grain-filling to maturity stages, when temperatures are below 30°C, the grain-filling rate increases with increasing average temperature, shortening the grain-filling period. When temperatures exceed 35°C during the initial grain-filling stage, grain-filling may be incomplete, leading to an increase in the number of unfilled grains, acceleration of the filling rate, and thus affecting the thousand-grain weight. In summary, the most vulnerable period of hybrid rice to high temperatures mainly occurs after the elongation-booting stage, especially during the heading and flowering stages where pollen dehiscence, pollen vitality, and stigma pollen numbers are most affected. Therefore, it is essential to avoid high temperatures during the flowering stage to improve the grain-setting rate in the production process of hybrid rice.

(5) *Characteristics of high-temperature tolerance adaptation in super hybrid rice.*

a) Atmospheric temperature. Rice is a thermophile, and temperature is one of the main climatic factors affecting its growth, development, and yield. In multiple-cropping areas where double-cropping rice is grown, the atmospheric temperature must meet the basic requirements for one season of rice growth. The accumulated temperature should be 2,000°C to 3,700°C per year, and effective accumulated temperature should occur in 110 to 200 days. The favorable atmospheric temperature conditions are the accumulated temperature between 5,800°C to 9,300°C per year and effective accumulated temperature time over 260 days per year provided that other conditions such as sunlight, precipitation, and soil fertility are adequate.
b) Atmospheric humidity. Atmospheric humidity is one of the ecological factors in rice growth and yield. Numerous studies have shown a significant correlation between atmospheric humidity and photosynthetic rate. Although its impact on grain-filling speed is not significant, excessive relative humidity within the population hampers grain-filling, which will cause diseases such as sheath blight and affect the duration of pollen viability during flowering, thereby influencing fertilization and grain-setting rates. Research on the effects of air temperature and relative humidity on grain-filling and empty husk rate from the late heading to maturity stage in late-season rice suggests that relative humidity has a minor impact on the grain-filling rate of super hybrid late-season rice. The strongest photosynthetic capacity of super hybrid rice is observed at a relative humidity of 50% to 60%, while the minimum relative humidity during dry, hot winds is ≤30%.
c) Interaction between atmospheric temperature and humidity. Atmospheric temperature and humidity are interlinked and act together. High temperature and humidity have an adverse effect on pollen cracking and pollen dispersal; high humidity causes higher surface temperature of rice, which in turn affects pollen activity. High temperatures (30°C to 35°C) and high humidity (relative humidity of the air 85% to 90%) during the heading stage of super hybrid rice will significantly increase the grain infertility rate. High temperature and high humidity conditions are not conducive to the opening of rice pores and temperature regulation of rice body tissue, leading to increased actual heat damage to rice, while moderate air humidity is conducive to lowering the temperature of the panicle and improving the nurturing ability of small panicles.

   When super hybrid rice is hot during the heading stage if it is subject to high humidity or drought, it will cause the infertility rate to increase dramatically; high humidity will cause lower air-saturated vapor pressure, stomatal conductance, and evaporation rate. Then with a higher leaf temperature, flower organs are severely damaged, pollen count is scarce, loose powder, and poor extensibility. Humidity has a direct impact on anther cracking. Within a certain range, the lower the humidity, the better the anther cracking performance; humidity has a greater impact on high-temperature hazards, and high or low humidity will aggravate hot disasters of high temperature. High temperatures and low humidity will exacerbate the adverse effects of droughts on the growth of super hybrid rice.

d) Specific measures for super hybrid rice to resist high temperatures. First, cultivate heat-resistant super hybrid rice varieties. Cultivating heat-resistant varieties is an effective way to solve the inadaptability of super hybrid rice to high temperatures. There are also some achievements in breeding. Some heat-resistant varieties have been successfully cultivated, but their agronomic characteristics are not very good, and the yield is not very high, so further research in this area needs to be carried out. For example, super hybrid rice with good heat resistance does not necessarily have the greatest potential yield under normal temperature conditions, but under high-temperature conditions, its yield stability is the best. Second, avoid frequent high temperatures at the heading and flowering stages. When the highest temperature was above or equal to 35°C for three consecutive days during the flowering period of the harvest spike, the production of super hybrid rice Xiangliangyou 900 was reduced by 360 kg/ha. From the analysis of the yield structure, the high-temperature index increased, the weight of effective panicles and thousand-grain both decreased, and the air-drying rate continued to increase. Therefore, the harvest period for planting super hybrid rice in the Yangtze River Basin was avoided from July 13 to August 21. At this time, hot weather was frequent, and the frequency and intensity of extreme heat were high, which had a serious impact on high and stable yield. Third, be aware of the layout of super hybrid rice varieties. In the production of super hybrid rice, varieties resistant to high temperatures should be deployed in the middle to late stages of super hybrid rice growth, especially in areas prone to high temperatures and droughts during the booting and heading-flowering periods, so as to reduce the harm of high temperatures to super hybrid rice production and stabilize the area and yield of super hybrid rice. On the other hand, some varieties that are not resistant to high temperatures have higher yields and should be placed in regions with particularly high temperatures to make full use of their yield potential. Therefore, different regions have different climates, and selecting varieties for cultivation and promotion cannot meet the requirements of uniform standards. Instead, they should start from the perspective of scientific analysis, adapt to local conditions, make full use of the environmental resource advantages of each region, give full play to the maximum production potential of varieties, and further increase the production of super hybrid rice.

6. Classification of cold damage and its impact on super hybrid rice

(1) *The definition of cold damage.* Cold damage refers to the phenomenon that during the growing season of rice, the temperature drops below its optimum growth temperature and above 0°C, which leads to stagnation of rice growth or stunting of growth.

(2) *Classification and hazards of cold damage.* The types of cold damage to rice can be divided into delayed cold damage, obstacle-type cold damage, and mixed cold damage.

a) Delayed cold damage refers to a delayed growth period caused by a long-term cold condition that occurs during rice's nutrient growth period. This type of cold damage generally occurs during the sowing and seedling period.
b) Obstacle-type cold damage occurs when abnormally low temperatures occur for a short period of time during the reproductive growth period (the differentiation of the genital organs until the ears bloom), which will disrupt the physiological function of the rice genital organs, causing infertility, grains, and hollowing of the flowers. This type of cold damage generally occurs during the heading period.
c) Mixed cold damage means that during the entire growth period of rice, the above two types of cold damage occur one after another.

The middle and lower reaches of the Yangtze River are in the subtropical monsoon climate zone. In spring, rice is susceptible to cold damage due to the influence of high-latitude atmospheric circulation and the western Pacific subtropical high pressure. During the planting period from March to April, cold air masses from the north frequently interact with warm and humid air masses from the south. When these air masses converge, cold and rainy weather forms. If cold and rainy weather persists for several days and the daily average temperature drops to 12°C or below, the seeds of super hybrid rice will rot, which is commonly known as "late spring coldness." This delayed cold damage will not only waste seeds but also lead to later replanting, resulting in a delayed growth period. In autumn from September to November, as the northern hemisphere gradually cools down, there is a rapid change in atmospheric circulation and climate. The northern cold air masses frequently invade China, and once strong cold air masses invade the south of the Yangtze River, they will cause cold damage.

The period around the date of Cold Dew is often a critical time for the heading and flowering of late-maturing super hybrid rice varieties. Due to the southward intrusion of cold air masses, there can be reduced yields due to empty grains and thin grains, which is known as autumn cold damage for super hybrid rice. This cold air mass is a "Cold Dew wind" that will cause obstacle-type cold damage. It adversely affects the germination of pollen grains, elongation of pollen tubes, flowering and pollination, opening of pistillate flowers, and rupture of anthers. It affects the synchronous heading of rice plants, leading to an increase in empty grains in panicles, a decrease in grain-setting rate, and ultimately causing yield reduction or complete crop failure.

At the same time, the study also showed that low temperatures can inhibit the photosynthetic ability of leaves of super hybrid rice plants, and the longer the low temperature lasts, the stronger the inhibitory effect. The grouting rate, fruiting rate, and dry matter weight of super hybrid rice are all negatively correlated with the duration of low temperatures.

(3) *Evaluation of indicators of cold damage.* Cold damage occurs when rice is exposed to continuous low temperatures and poor lighting for a long time or strong low temperatures for a short time during the growing season (TABLE 8.3).

TABLE 8.3 Super hybrid rice low-temperature stress index

Growth stage	Low-temperature stress index (Sun Wen, 2008)	Damages (Chen Yan, 2021)	Symptoms
Seedling stage (30 days after sowing)	Average daily temperature ≤12°C for more than three consecutive days	The critical average daily temperature is 16°C–18°C. A decrease of 1°C in average temperature will extend booting by three to four days.	Grow slowly, resulting in seedling decay and seedling rot.
Booting stage (45 days before heading)	Average daily temperature ≤20°C for more than three consecutive days	The critical average daily temperature is 18°C. A decrease of 1°C in average temperature will cause a decrease in grain-setting rate by about 6%.	Reduction in the number of differentiated tillers and grains per panicle, with a large number of sterile grains.
Heading and flowering stage (20 days before and after heading)	Average daily temperature ≤22°C for more than three consecutive days	Critical average daily temperature is ≥20°C, it will significantly impact on yield.	Decrease in pollen germination rate, failure of anther dehiscence, reduction in angle of pistillate flower opening, affecting fertilization, increasing empty and shriveled grains, and incomplete grain maturation.

To determine the impact of the duration of extremely low temperatures on super hybrid rice production, different days of low temperatures during different growing periods were quantified into extreme low-temperature stress. That is, according to the extent of damage caused to super hybrid rice by the number of days the low temperature lasts, heat and cold damage can be divided into light (three to four days), intermediate (five to six days), heavy (seven to nine days), and extra heavy (more than nine days) (TABLE 8.4).

TABLE 8.4 Different consecutive days of extreme low temperature stress severity (Sun Wen, 2008)

Growth stage	Consecutive days of extreme low temperature/day							
	2	3	4	5	6	7	8	≥9
Seedling stage	—	1.0	1.5	2.0	2.5	3.0	3.5	4.0
Booting stage	1.0	1.04	1.08	1.12	1.17	1.21	1.26	1.31
Heading and flowering stage	—	1.0	1.04	1.08	1.12	1.71	1.21	1.26

The Yangtze River Basin is one of the main regions for rice cultivation in China. However, during the spring and autumn seasons, influenced by cold air from the north, the middle and lower reaches of the Yangtze River are prone to experiencing late spring coldness, resulting in seedling decay, and the Cold Dew wind in autumn, affecting heading and grain-filling. These are the two main types of cold damage affecting the growth and development of rice in southern China.

(4) *The impact of cold damage on yielding.* Late spring coldness and the Cold Dew wind pose serious threats to the growth of hybrid rice in the middle and lower reaches of the Yangtze River. For instance, late spring coldness can lead to seedling decay, causing substantial losses in rice seeds. Due to late spring coldness, the region suffered losses of 400 million kg of rice seeds in 1970 and 650 million kg in 1976. In 1971 and 1972, around 10% of rice fields in the middle and lower reaches of the Yangtze River were affected. The impact of Cold Dew wind on rice yields is also significant. In late September 1971, Jiangxi experienced seven consecutive days of northerly winds, with daily average temperatures dropping below 17°C. The result is 30% to 50% empty grain rates in multiple counties and cities, and exceeding 70% in some areas, even crop failure. In 2020, due to cold and cloudy weather in the middle and lower reaches of the Yangtze River, including Hunan, Jiangxi, Hubei, and Jiangsu provinces, empty rates ranged from 30% to 90%, with some areas experiencing no grain harvest at all. Due to climate change, severe cold damage in the middle and lower reaches of the Yangtze River has decreased in recent years, mainly manifesting as mild disasters. However, climate change has also led to more frequent extreme weather events, with significant temperature fluctuations during the growth period of hybrid rice. Short-term extremely low temperatures still pose a frequent and severe threat, with the frequency of rice cold damage between 2001 and 2009 being higher than that between 1991 and 2000. Furthermore, climate change has caused the northern boundary of the double-cropping rice planting area in the middle and lower reaches of the Yangtze River to move northward by nearly 300 km, increasing the risk of hybrid rice suffering from cold damage.

Cold damage to super hybrid rice has the characteristics of being sudden and cluster, and timely and effective cold protection measures can reduce losses caused by cold damage.

(5) *Specific measures for super hybrid rice to address cold damage.*

a) Select local super hybrid rice varieties. Choose varieties with strong cold resistance, mainly mid-maturing varieties.
b) Early planting timely. When the average daily soil temperature stabilizes above 12°C, early planting can be done timely. Based on the length of the grain-filling period of super hybrid rice varieties, determine the appropriate heading stage, then deduce the suitable transplanting period. This not only can defend against cold damage but also avoid premature heading, which would waste temperature resources in the later stage and lead to early aging, as well as reduce the probability of panicle blast and rice blast.
c) Control the use of nitrogen fertilizer and increase phosphorus, potassium, silicon, organic fertilizers, and foliar spray magnesium. In years with low temperatures, it is recommended to reduce nitrogen fertilizer by 20%–30%, with a reasonable ratio of nitrogen fertilizer between the pre- and post-heading stages (70%–80% as basal and tillering fertilizer), which is conducive to early maturity and avoids low temperatures. Increasing phosphorus, potassium, and silicon fertilizer can make rice plants robust and enhance resistance to stress, promoting not only post-transplanting regrowth but

also accelerating heading, flowering, and maturation. The addition of organic fertilizer not only improves soil quality but also promotes early maturity. Foliar spraying magnesium fertilizer during the grain-filling period can improve the quality of super hybrid rice, resulting in a better taste of rice.

d) Scientific field management. Deep water irrigation during panicle differentiation in the presence of low temperatures; application of growth regulators to promote early maturity, such as a mixture of paclobutrazol, potassium dihydrogen phosphate, and urea, has proven to be effective.

CHAPTER 9

Comprehensive Prevention and Control Techniques for Major Diseases, Insects and Weeds of Super Hybrid Rice

I. Disease Prevention and Control Techniques

1. Rice blast

Rice blast is a fungal disease and one of the three major diseases of rice in China. It is also the biggest threat to super hybrid rice. Rice blast occurs in different degrees in both southern and northern rice areas in China and in any stage of rice's growth period. The severity varies depending on the year and region. Severe affection occurs in areas that experience less sunlight and longer terms of fog and haze, such as mountainous areas and mild climate areas along rivers and coasts. The rice blast fungus overwinters on rice straw and grains through conidia and mycelia and is primarily transmitted by infected seeds. When temperatures are favorable in the following year, the produced conidia spread through wind, rain, or air currents. The symptoms of the disease vary depending on the time and infected location. Therefore, based on the location, the disease is classified as follows: seedling blast occurs before the three-leaf stage of seedlings, leaf blast occurs on rice leaves after the three-leaf stage presenting disease spots, and node blast usually occurs after heading, presenting dark brown spots around the nodes (figure 9.1). Neck blast occurs at the panicle neck, starting as dark brown and later turning black, leading to the formation of whiteheads (figure 9.2).

Figure 9.1 Leaf blast

Figure 9.2 Neck blast

Grain blast occurs on the glumes of the grains, producing brown oval or irregular spots. The main reasons for the variability in disease severity among rice fields are the cultivation management practices (such as fertilization and irrigation) and the disease resistance of the varieties. The integrated disease management strategy is based on the selection and promotion of high-yielding, disease-resistant varieties (combinations), focusing on fertility and water management practices to achieve high yields, eliminating overwintering sources of the pathogen as much as possible, timely chemical control, and applying a "focus on both ends, smart management in the middle" control strategy. The integrated control technologies and measures are as follows:

(1) *Promotion of disease-resistant varieties.* Choosing varieties (combinations) with strong disease resistance is the most economical and effective control measure. The disease-resistant varieties should be well arranged according to the physiological races of the rice blast pathogen. Attention should also be paid to the timely replacement of varieties (combinations) in production. Varieties (combinations) should be replaced that are highly susceptible or have poor resistance. Adjusting varieties (combinations), on the one hand, can avoid the disease season; for example, in the early-season rice cropping areas of southern China, the heading period coincides with the plum rain season, which is conducive to the occurrence of neck blasts. Cultivation of late-maturing varieties in infected areas can avoid the climatic conditions favorable for neck blasts. On the other hand, adjusting varieties can prevent the pathogen's physiological races from changing due to the monoculture of a single variety. The promotion and application of different varieties (combinations) of super hybrid rice should be based on understanding and mastering the level of resistance to blast disease, and large-scale planting of the same variety should be avoided.

(2) *Eliminate overwintering sources of the pathogen.* Do not use diseased seeds, and deal with diseased grains and straws promptly. During harvest, the grains and straws from diseased fields should be piled separately and dealt with before the spring planting season. Be careful not to use

diseased straws for sprouting seeds or tying seedlings. When returning diseased straws to the field, it should be deeply turned and thoroughly decomposed. Straws used for compost should be fully matured before application.

(3) *Seed disinfection treatment.* Rice seeds should be selected from fields with no or mild disease. Seeds carrying the pathogen are one of the initial sources of seedling blast and leaf blast; thus, seeds need strict disinfection and sterilization. Generally, we choose coated seeds. As for the uncoated, disinfection is a must. For seed disinfection, you may choose any one of the following methods: Use a chlorine bleach solution to soak the seeds in a diluted solution of trichloroisocyanuric acid (trichloroisocyanuric acid:water = 1:300) for 12 hours, then rinse the seeds thoroughly with clean water before soaking them again or directly germinating them; or use seed soaking agent to soak every 6 kg of seeds in a mixture of 2 ml of the agent and 10 kg of water for 48 to 72 hours, after which the soaked seeds can either be germinated or directly sown; or use a 40% prochloraz or 50% carbendazim WP diluted to 1,000 times, and soak for two days, then lift and drain the seeds, and then rinse off the fungicide with clean water and germinate; or soak the seeds in a 20% triadimefon WP solution diluted to 500 times for 48 hours, which can be replaced with a triadimefon solution at 0.1% of the seed weight, or a 25% imazalil solution diluted to 2,000 times for seed disinfection.

(4) *Strengthen healthy cultivation.* Sowing an appropriate amount and cultivating thick, vigorous, disease-free seedlings are key to controlling seedling and leaf blast. Scientific management of fertilization and irrigation is an important measure in the integrated management of rice blast disease. It is important to pay attention to reasonable fertilization and the combination of the "NPK." The principles of fertilization are as follows: apply enough base fertilizer, apply topdressing early and in the middle and later stages, apply fertilizer cleverly based on the condition of the seedlings, weather, and field, and increase the application of phosphorus and potassium fertilizers. Appropriately apply silicon, zinc, and boron fertilizers. Do not apply nitrogen fertilizer singly. Apply panicle fertilizer at the right time. In areas where the disease is common and fields that are prone to disease, panicle fertilizer should not be applied or applied cautiously. Water management must be closely coordinated with fertilization, implementing scientific and reasonable drainage and irrigation, adjusting fertilization with water management, frequently irrigating with shallow water, and combining with field sunning to achieve both promotion and control. Adopt different water management methods in different growth stages of rice. Sun-drying the fields in a timely manner at the end of tillering can enhance the plant's disease resistance and control the development of leaf blight. Irrigating with shallow water during the heading stage to meet the needs of the flowering period. Moistening irrigation during the milk ripening stage and alternating dry and wet conditions during the maturing stage, which is conducive to producing fully ripe golden stalks and reducing disease incidence.

(5) *Conduct timely chemical control.* Based on forecasting and field surveys, it is essential to timely application of chemicals to protect rice plants during their susceptible stages and varieties prone to infection. In areas where rice blast disease is prevalent, chemical control aiming at suppressing seedling blast, leaf blast, and harshly treating panicle neck blast should be adopted. In these areas, it is recommended to apply chemicals at the three-to-four-leaf stage of the seedlings or five to seven days before transplanting to prevent seedling blast. During the tillering stage, focus on controlling rice fields that become centers of leaf blast outbreaks. Fields with a disease leaf rate of more than 2% at the end of the booting stage or a flag leaf disease rate of more than 1% need timely chemical treatment. The critical period for chemical control of panicle neck blast is during the necking and

heading stages (when the rate of necking is 10%, and the heading rate is around 5.0%), and if necessary, chemicals can be applied again during the later heading stage. Common chemical dosages per *mu* include 9% pyraclostrobin microcapsule suspension at 1,000 times dilution, 22% kasugamycin and tricyclazole suspension at 500 times dilution, 20% triadimefon WP at 100 g, 75% triadimefon WP at 40 g, 40% isoprothiolane at 100 ml, 25% prochloraz at 100 ml, or 2% kasugamycin solution at 100 ml. Other agents such as polyoxin, picoxystrobin, trifloxystrobin · tebuconazole are also used.

2. Rice sheath blight

Rice sheath blight, commonly known as "flower foot disease" in China, is a fungal disease that frequently occurs and causes severe damage to super rice. This disease is characterized by its wide occurrence, high frequency of outbreaks, serious harm, and significant losses. The pathogen of sheath blight primarily overwinters as sclerotia in the soil of rice fields or on the remains of diseased grasses and weeds. When the temperature becomes suitable, the sclerotia germinate, producing mycelia that invade the leaf sheath later. The disease can occur from the seedling stage to the heading stage, generally during the peak and late stage of tillering, and heading stages, with the period around heading being the most severe. The tillering and booting stages are the most susceptible to infection. The disease mainly affects the leaf sheaths and leaves of rice plants and damages the panicles or the stems in severe cases. Initial symptoms are small dark green water-soaked spots on leaf sheaths near water level, and gradually expanded into oval cloud-like lesions. The occurrence and prevalence of sheath blight are influenced by several factors, including the amount of pathogen, variety resistance, planting density, climate conditions, and fertilization and water management. However, the microclimate in the field and the growth stage of the rice plants are the main factors affecting the severity. Sheath blight is a disease of high temperature and high humidity, and it is also related to vigorous growth with lush green leaves due to excessive nitrogen fertilizer application. Super hybrid rice, with its thicker stems and more abundant leaves, tends to grow vigorously, leading to a closed canopy and increased humidity in the fields. The occurrence of heavy rain can further exacerbate the disease. For some paddy fields that are irrigated for a long time, late and uneven application of nitrogen fertilizer can cause excessive elongation. Adding to high density that means lack of ventilation and light, the fields that are insufficiently exposed to sunlight are favorable for the development and spread of sheath blight (figure 9.3). The integrated pest management strategy includes eradicating or reducing the source of pathogens, enhancing fertilizer and water management, applying formula fertilizers, sunning fields at appropriate times, and timely applying pesticides based on field disease conditions and control indicators. The comprehensive approaches are as follows:

(1) *Eliminate the sclerotia and reduce the source of infection.* Sclerotia floating on the water surface should be collected during water management and field harrowing activities. It is advisable to salvage them in large continuous patches as much as possible and consistently every season before and after planting. The collected sclerotia should be removed from the fields and either deeply buried or burned, which aims to minimize the source of infection in the field. Additionally, weeds around the field should be removed, as they can harbor the fungus.

Figure 9.3 Sheath blight (early stage, middle stage, late stage)

(2) *Select resistant varieties.* Although no varieties immune or highly resistant to sheath blight have been identified, there are differences in resistance and susceptibility among different varieties (combinations). Generally, broad-leaf varieties are more susceptible than narrow-leaf varieties. Super hybrid rice varieties tend to have a stronger tolerance to the disease due to their vigorous growth traits.

(3) *Enhance healthy cultivation practices and strengthen fertilizer and water management.* Adopt a reasonable planting density. Adjust row spacing according to local conditions to facilitate wide and narrow row planting. This improves ventilation and light penetration within the rice crop, reduces field humidity, and subsequently lessens disease severity. Apply sufficient base fertilizer early and follow up with topdressing. Combine nitrogen fertilizers with phosphorus and potassium fertilizers and increase the application of silicon fertilizers. Avoid the excessive use of nitrogen fertilizers. Promote the use of balanced fertilization techniques to ensure the rice does not exhibit excessive nutrient growth at any stage. As for water management, follow the "shallow in early stage, sunning in mid-stage, and moist in late stage" principle. Ensure shallow-water irrigation for tillering, exposed fields for seedlings, and sun-drying for root growth. Apply heavy sun-drying on fertile fields and moderate sun-drying on lean fields. Maintain moisture for panicle development, and timely cut off water supply to prevent premature aging of the crop. Particularly, sun-drying the field in the mid-stage is crucial for promoting sturdy growth, using water management to control the disease, and enhancing disease resistance. Note that do not prolong deep watering or excessive sun-drying.

(4) *Scientific, rational, and timely application of fungicides.* The best time for applying fungicides to control Sheath blight is from the late tillering stage to the early panicle initiation stage of rice growth. Generally, the disease rate reaches 15.0% to 20% at the end of the tillering stage and 20% to 30% at the early stage of booting. For super hybrid rice, when the disease rate exceeds 30% as the leaf beneath the flag leaf (the second-to-last leaf) fully emerges, applying fungicides again at this stage shows the best results. Consider a third application based on actual field conditions. This could be combined with the control of pests like the rice leaf folder, planthoppers, and other disease insects. Especially in high-temperature and high-humidity conditions that often coincide with rainy weather, it might be necessary to apply fungicides two to three times consecutively with an interval of 10 to 15 days between applications. As for recommended dosages, benzothiadiazole formulations should be 30 ml of 325 g/l concentration with 60 ml diniconazole formulations of a 19% concentration;

alternatively, validamycin formulations should be 100 g of 5% water-SP, or 200 ml of 5% aqueous solution, or 100 ml of 10% aqueous solution, or 30 g of 20% water-SP. Mixing validamycin with carbendazim can enhance effectiveness. Other combinations are 200 ml/*mu* of 12.5% validamycin aqueous solution or equivalents in other formulations, 20 ml/*mu* of 30% propiconazole emulsifiable oil, or 40 ml of 10% hexaconazole emulsifiable oil. Additionally, there are other fungicides such as propiconazole, fluquinconazole, tebuconazole, prochloraz, and thifluzamide. Any one of these fungicides above can be selected for use, diluted with water, and applied at a rate of 50 to 60 kg/*mu*. It can be applied through manual spraying or using unmanned aerial vehicles (UAVs) at a rate of 50 to 60 kg of water per *mu* to ensure even coverage, particularly targeting the middle and lower parts of the rice plants.

3. False smut

Rice false smut, also known as green smut or grain flower disease, colloquially referred to as "harvest fruit," is a fungal disease. This disease affects super hybrid rice with a higher incidence rate compared to conventional rice. The pathogen overwinters in the soil as sclerotia or teliospores on seeds and germinates to produce conidia when the climate becomes favorable in the following year. These conidia, spread by air currents, pose a threat to the rice panicle from flowering to the milky stage. As the fungus grows inside, the initially infected grain will slightly open and reveal small yellow-green protrusions that gradually enlarge and envelop the glume, forming false smuts. These smuts are several times larger than the grain, nearly spherical, yellow-green or dark green, and cracked, releasing toxic dark green powder. Usually, one to several smuts can be seen in a single panicle, and in severe cases, a dozen or more in a panicle. Rice false smut not only destroys the grains but also consumes the nutrients of the entire affected panicle, leading to unfilled grains. As the number of diseased grains increases, the rate of empty grains rises, and the thousand-grain weight drops, significantly affecting yield and causing severe deterioration in rice quality. Rice false smut can occur from the heading stage to maturity, with the highest susceptibility at the late booting stage. Climatic conditions, especially rainfall and temperature, are crucial factors affecting the occurrence and damage caused by rice false smut. During the booting to heading stage, the fungus grows most suitably in high temperatures and humidity, while prolonged low temperatures with little sunlight and abundant rain can weaken the rice's resistance to the disease. During the flowering period, continuous rain, low temperatures, or waterlogged conditions in the field can all contribute significantly to the outbreak of rice false smut. Additionally, the increased use of chemical fertilizers, particularly excessive nitrogen fertilizers and panicle fertilizers, can result in lush, tender, green growth of rice after heading, causing the plants to mature late and be more susceptible to disease. The incidence of the disease is generally higher in super hybrid rice and large-panicle hybrid types (figure 9.4).

The recommended approaches are as follows:

(1) *Select high-yielding disease-resistant varieties (combinations).* Different rice varieties show varying levels of resistance to diseases. Generally, early-maturing varieties with loose panicles have a lower incidence of disease, while late-maturing varieties with large or dense panicles have a higher

incidence. Hybrid rice combinations with a longer booting stage tend to be more susceptible than those with a shorter booting stage.

(2) *Use disease-free seeds and conduct seed disinfection.* Before sowing, use salt water to select seeds, eliminating diseased grains. Seeds should be soaked in warm water at 57°C for 10 minutes, then washed and germinated for sowing. Using 0.5 kg of quicklime diluted in 50 kg of water for soaking 30–35 kg of seeds for about 72 hours can be effective. Alternatively, soaking seeds in a 1,000-fold dilution of 10% bactericide "401" for 48 hours can sterilize and promote germination. Another method is to mix 1.0 to 1.5 g of 15% triadimefon powder with 1 kg of rice seeds for 24 to 48 hours, which can be directly sown without the need for further germination. Mixing 50 ml of a 3% suspension seed dressing of benomyl per *mu* has also shown favorable results. Additionally, applying 50 ml/*mu* of 3% difenoconazole FS seed dressing can provide good results.

Figure 9.4 False smut (light severity, heavy severity)

(3) *Strengthen field management.* Early detection and removal of infected grains in the field should be practiced, with the grains being taken out of the field and burned. After harvesting severely infected fields, deep plowing and burying of sclerotia should be conducted. The fields should be maintained clean. Before sowing rice, it is crucial to remove any remaining diseased material and pathogens from the field to reduce the source of infection.

(4) *Insist on rational fertilization.* Apply organic fertilizers generously and use nitrogen, phosphorus, and potassium fertilizers in combination to avoid excessive or late application of nitrogen fertilizers. Fertilizers should be applied as base fertilizer and topdressing early, with tillering and panicle fertilizers each accounting for 1/3, and avoid excessive application of panicle fertilizer. Scientific application of foliar fertilizers to supplement nutrients is recommended. During the heading stage, for each *mu*, 150 g of monopotassium phosphate and 100 g of zinc sulfate can be dissolved in 50 kg of water for spraying. Additionally, a 0.1% to 0.2% solution of borax can be sprayed.

(5) *Strengthen monitoring and forecasting.* According to research reports, the rice false smut pathogen infected late-maturing rice generally in early to mid-September. The rain temperature coefficient during this period significantly affects the number of diseased grains in early to mid-October. Many rainy days in early to mid-September, adding with temperatures between 25°C to 30°C, is conducive to the occurrence of the disease. Note to prepare for prevention and control.

(6) *Chemical control.* For the prevention and treatment of rice false smut, the first application of chemical pesticides is five to seven days before the necking period, and the second application is during the necking period, which can have a good effect. Commonly used pesticides include 20 ml of 30% tebuconazole emulsifiable concentrate (EC) per *mu*; 100 to 120 g of 15.5% prothioconazole WP; 80 to 100 g of 20% jinggangmycin WP; 80 g of 15% triadimefon WP; 40 ml of 10% imidacloprid EC; 60 ml of 23% difenoconazole EC. Each should be mixed with 50 kg of water for conventional spraying on rice plants or application using agricultural UAVs. Additionally, triazole fungicides such as propiconazole, fluquinconazole, tebuconazole, triazolone, and oxymatrine have shown good effectiveness.

4. Bacterial leaf blight and bacterial stripe disease

Rice bacterial leaf blight is a bacterial disease and one of the major diseases of rice. In the 1980s, it caused a widespread epidemic in the rice-growing areas of southern China. However, as the resistance of varieties improved, the severity of the disease gradually reduced over the years, but in recent years, there has been a resurgence trend of super hybrid rice in coastal areas. The overwintering pathogens, mainly on rice seeds and straw, are the primary sources of initial infection in the following year. The bacteria enter through water pores and wounds on the leaves after sowing or transplanting and spread through wind, rain, and flowing water. Rice can be affected throughout its entire growth period, mainly damaging the leaves. The lesions start at the leaf tips and edges, later developing into streaks along the leaf edges or midrib in a wavy pattern, with the lesions being yellowish-white. There is a clear boundary between the diseased and healthy parts. In severe cases, the lesions turn gray-white, giving a withered white appearance from a distance. When the field humidity is high, yellow bead-like bacterial ooze can be seen on the diseased leaves, which dries up into hard grains that easily fall off. Typical symptoms observed in mature plants include leaf scorch, acute type, midrib type, wilting type, and chlorosis type. Rice is most susceptible during the heading stage, followed by the tillering stage (figure 9.5).

Additionally, bacterial stripe disease (also known as stripe disease) is a bacterial disease. Initial symptoms are red striations at the leaf tips, which quickly spread along the leaf veins to connect into larger areas under continuous gloomy, rainy, and windy weather. When it worsens, the entire field's leaves will turn red or yellow-brown, accompanied by yellow bacterial oozes (figure 9.6).

The occurrence and epidemic of bacterial leaf blight and stripe disease are closely related to climatic factors, water and fertilizer management, and the resistance of varieties. Particularly, heavy storms, floods, deep irrigation, relay irrigation, and excessive nitrogen fertilizer create favorable conditions for its rapid spread and potential disaster. The integrated control strategy for these two diseases includes planting resistant varieties and disease-free seeds to prevent seedling diseases, strengthening water and fertilizer management, and applying pesticides timely. The approaches are as follows:

(1) *Selection of disease-resistant or disease-free varieties (combinations).* There is a significant variation in resistance to bacterial blight among different rice varieties. Generally, disease-resistant varieties offer better protection against the occurrence and harm of bacterial blight, and their resis-

Figure 9.5 Symptoms of bacterial leaf blight in the field

Figure 9.6 Symptoms of bacterial stripe disease in the field

tance is relatively stable. In areas where the disease is prevalent annually, it is advisable to choose two to three main disease-resistant varieties suitable for local cultivation. Moreover, it is best to produce seeds of super hybrid rice in disease-free areas to prevent the introduction of pathogens.

(2) *Strict seed disinfection and proper handling of diseased straws.* Seeds carrying bacteria must be treated with chemicals. The following methods can be adapted for seed disinfection: Carbendazim, a 300-fold solution of 40% hypochlorous acid, a 500-fold solution of 85% trichloroisocyanuric acid, a 500-fold solution of 20% thiophanate methyl WP, a 12-hour soak in a 70% allyphl-sulfide solution, or a 2,000-fold solution of 10% blasticidin-S. Then, rinse the seeds well, germinate, and sow. Prevent waterlogging in seedling fields and cultivate healthy, disease-free strong seedlings. Rice straw residues should be dealt with promptly; avoid using diseased straws to bind seedlings, cover seedlings, block field entrances, etc. Select seedling fields that are sheltered from the wind, face the sun, are situated on higher ground, and have convenient drainage and irrigation to prevent flooding of fields by heavy waters.

(3) *Strengthen fertilizer and water management and promote healthy cultivation practices.* Improve the irrigation and drainage system, implement separate systems for irrigation and drainage, avoid deep watering, prohibit sequential and indiscriminate irrigation to prevent waterlogging. Apply sufficient base fertilizer and early topdressing, fertilize scientifically according to changes in leaf color and formulas, adhere to the balanced application of nitrogen, phosphorus, potassium, and trace elements to enhance the plant's disease resistance, ensuring healthy and stable growth of the seedlings.

(4) *Applying chemical treatments in key periods.* The key to chemical treatment is early detection and early treatment. The focus in disease-prone areas is to spray the seedlings during the seedling stage to protect them. As for late maturing field, spray before irrigation and seedlings transplantation. During the cropping season, isolate diseased plants, and disease centers. The principle of "treat a small area upon finding a single affected spot, treat the entire field upon finding a larger area" should be applied. Immediate chemical treatment is required in the initial period of the tillering stage and the early stage of booting, especially when acute disease spots appear in the field and the weather conditions favor disease development. Especially attention should be given to the infected fields and neighboring rice fields after big storms, and flooded fields, which will effectively control the spread of the disease. For prevention and control, the main chemicals and their dosages per *mu* are 100 g of 20% kasugamycin (Yeqing Shuang) WP, 200 g of 25% validamycin (Yu 7802) WP, 100 g of 70% ka-

sugamycin hydrochloride (Shaku Qing) JG, 100 ml of 20% oxolinic acid copper (osencopper) SC, or 80 g of 90% streptocide SP. Alternatively, a 600-fold solution of trichloroisocyanuric acid, a 700-fold solution of Jinmou Cuoxin oxime ketone, or a 1,300-fold solution of acrobat benziothiazolinone can be used. Additionally, 3% zhongshengmycin, 24% streptomycin, 2% ningnanmycin, bacillus subtilis, and other biological agents are available. Any of the above agents can be chosen and used in rotation, diluted in 50–60 kg of water for conventional foliar spraying, or applied using an agricultural UAV for ultra-low-volume spraying. The treatment should be repeated every 7–10 days, for a total of 2–3 applications.

Please refer to the preventive and control techniques for bacterial stripe disease (stripe disease) to that of bacterial leaf blight, for they are similar.

II. Pest Control Techniques

1. Rice stem borers

Rice stem borers commonly known as "heart borers," belong to the family of moths in the order Lepidoptera. Super hybrid rice has robust stems, large pith cavities, dark green leaves, high nutrient content, high starch content, high soluble sugar content, and reduced silica content. All these are favorable conditions for the widespread occurrence of stem borers, resulting in broad infestation areas, high frequency of outbreaks, high pest density, and severe damage. Stem borers have become a major pest in the rice-growing regions of southern China since the 1980s, following the development of hybrid rice. The larvae, which are more cold-resistant, overwinter inside rice in rice stubble, straw, or wild rice stems. This pest damaged early-maturing rice before, and now it also damages both mid-season and late-maturing rice crops. Stem borers damage rice primarily through larval boring, causing withered sheath and seedlings in the tillering stage (figure 9.7), dead booting panicles during the booting stage, whitehead panicles during the heading stage (figure 9.8), and insect-damaged plants or semi-withered panicles during the milk ripening stage. If boreholes can be seen in the infested stems, there often are multiple larvae and much frass within a single stem. Adult stem borers are attracted to light and tender green plants, with each female laying 2–3 egg masses, each containing 50–80 eggs. After hatching, the larvae initially cluster inside the leaf sheaths, causing damage by boring into the leaf sheath tissue, which leads to dead sheaths. As they grow into the second instar, they disperse and bore into the interior of the rice plant, causing dead buds or whiteheads.

To control the rice stem borer, adopt a strategy that is "firmly control the first generation, manage the second and third generations, and selectively target the fourth generation." This involves the promotion and application of green control technologies such as pheromone traps, trichogramma (often referred to by its common name "red-eyed wasp"), and insecticidal lamps to minimize damage from the pest. The approaches are as follows:

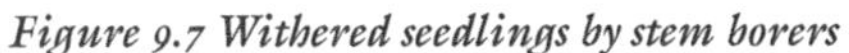

Figure 9.7 Withered seedlings by stem borers

Figure 9.8 Whitehead by stem borers

(1) *Submerging pupae in deep water to reduce the base number of overwintering pests.* Idle fields in winter should be plowed and flooded before winter or deeply flooded for three to five days in early April of the following year; in lake regions, non-tillage fields can be long-term flooded after spring; and fields intended for green manure can be flooded for two to three days in early April. Flooding fields with deep water can drown most of the overwintering larvae and pupae. Early flooding and plowing of seedling throwing fields, combined with spring plowing that turns rice stubble and straw into the soil, can help eliminate overwintering pests on rice straw, thereby effectively reducing the pest population.

(2) *Rearranging rice planting patterns to avoid pest infestation.* Promote rice cultivation techniques that reduce pest infestation. By transforming the coexistence of single and double-cropping rice into large-scale contiguous planting of either double-cropped or single-cropped rice, pest sources can be effectively cut off. This minimizes the presence of "bridge fields" that facilitate the reproduction of the second-generation rice stem borer. Properly adjust and reasonably match early rice varieties of early, medium, and late maturity, prioritizing early to medium-maturing varieties while reducing late-maturing varieties. Timely transplantation can avoid the damage period of the first-generation rice stem borer and the critical growth stage of rice, reducing both the number of the first-generation pest and the annual base number of pests, achieving avoid pests through cultivation.

(3) *Select varieties with strong resistance.* The resistance of rice varieties (combinations) to the rice stem borer significantly affects the severity of infestation. Varieties (combinations) with better resistance or tolerance generally have such features: thicker stem walls, smaller medullary cavities, smaller distances between vascular bundles, and smaller air spaces in the leaf sheaths. The number of silica cells in the stems and leaf sheaths is higher, and the chlorophyll content is lower. Also, reduce and eliminate susceptible varieties (combinations). Super hybrid rice may be more susceptible in the early stages but becomes less susceptible later due to its thicker stems and stem walls, making it harder for pests to invade.

(4) *Protect and utilize natural enemies.* Various predatory spiders in the fields can prey on the hatched larvae of the rice stem borer, especially those from the wolf spider family, which have strong predatory capabilities. A single pirate spider can consume all the larvae hatched from a cluster of rice stem borer eggs in one day. Late-season rice fields have numerous natural enemies, including

major predators such as spiders, ground beetles, ladybugs, and other predatory natural enemies. Main parasitic natural enemies include Chinese and Australian parasitoid wasp, Cotesia ruficrus, Charops bicolor, Brachymeria lasus, and tachina fly. Where possible, artificial release of parasitoid wasp in rice fields can be effective. When using pesticides, select types and formulations that are less harmful to natural enemies. Application methods needs to be perfected to protecting natural enemies in the field.

(5) *Use of insect pheromones (sex attractants).* Utilize sex attractants for the rice stem borer to conduct field trapping, killing adults and disrupting the mating of male and female moths. Commonly what we used in production are water pan traps. Make sure that the edge of the pan is always about 20 cm higher than the rice plant. The lure should be 0.5 to 1.0 cm above the water surface in the pan, with 0.3% detergent added to the water. Water should be added every evening to maintain the level, change the water and detergent every 10 days, and replace the lure every 20 to 30 days to achieve harmless control. Recently, white cylindrical traps and cage traps have been widely used in production; after installing the lure, they can be directly inserted into the fields, placing 1–2 per *mu*.

(6) *Use of moth-attracting insecticidal lamps for light trapping.* Utilize the phototactic behavior of the rice stem borer to kill adult moths, especially during peak moth emergence periods, yielding effective results. Each light can typically kill more than 200 adult stem borers per night. Each lamp can cover three to four ha of rice fields, reducing egg-laying by about 70% in the controlled area, resulting in fewer pests and less damage. Lights are generally turned on in early May and turned off in mid-October. This method, one-time investment for reusable, is economic and non-toxic effectiveness. Currently, products that are widely used includes the Henan Jiaduo frequency-vibration insecticidal lamp and the Hunan Shenbu brand fan-suction harm-benefit separation insecticidal lamp.

(7) *Timely chemical control.* For the rice stem borer, inspect egg mass hatching progress to determine the optimal timing for control, and inspect seedling conditions, egg mass density, and the rate of dead sheaths to identify fields requiring treatment. Understand pest conditions, select the right pesticides, and grasp the key period from peak larval hatching to the low-age larva stage. The chemical control strategy is "firmly treating the first generation, controlling the second and third generations, and selectively targeting the fourth generation while considering both earlier and peak times." Control indicators and optimal timing are as follows: Typically, fields with more than 50 egg masses per *mu* or a dead sheath rate of more than 3.0% per field are considered for treatment. In general years, apply pesticides once two to three days after the peak hatching of larvae. And in years of severe infestations, apply once at the peak of hatching and again five to seven days later. Apply pesticides at the early stage of rice panicle initiation (about 10% panicle initiation) can get best effect in controlling whiteheads. Commonly used pesticides and application methods in recent years are 20% chlorantraniliprole at 15 ml/*mu*, 5% abamectin at 100 to 150 ml/*mu*, 5.7% emamectin benzoate (emamectin salt) at 80 to 100 g/*mu*, and 10% methoxyfenozide at 100 ml/*mu*. For the various pesticides mentioned above, any one of them can be selected for application based on actual conditions. The recommended dosage per *mu* should be mixed with 50 to 60 kg of water for manual spraying or mixed with 6 to 8 kg of water for motorized spraying. Alternatively, agricultural UAVs can be used for pesticide application. Keep the layer of water in field at three to five cm, and no irrigation for five to seven days after application.

2. Rice leaf roller

The rice leaf roller, also known as the rice green caterpillar or white leaf caterpillar, belongs to the moth family of the Lepidoptera order. It is a migratory pest that severely affects rice-producing areas in southern China, including Hunan, Hubei, Jiangxi, the northern parts of Guangdong and Guangxi, and the southern part of Zhejiang, with five to six generations occurring annually. The adult moths display migratory, phototactic (attracted to light), and chlorotactic (attracted to fresh green) behaviors. The first generation usually causes light damage, with the second generation's larvae harming the early rice during the booting and heading stage in mid to late June. The first-generation larvae damage single-cropping rice and mid-season rice from late July to early August, while the fourth-generation larvae affect late-season double-cropping rice in early to mid-September. The larvae of the rice leaf roller damage the rice leaves by spinning silk and rolling the leaf longitudinally into a shelter, where they hide and feed on the leaf tissue, leaving only the epidermis and forming white streaks. When infestations are severe, numerous leaf rolls can be seen throughout the fields (figure 9.9).

The growth of rice during the tillering stage is inhibited by the damage, and during the heading stage, the functional leaves are damaged, affecting the grain-setting rate and the thousand-grain weight. The occurrence of the rice leaf roller is closely related to climatic conditions and the ecological environment. Conditions such as an average temperature of 25°C, frequent rainfall, high precipitation, and relative humidity above 80% are conducive to the proliferation of this pest. In areas where early, medium, and late rice are intercropped and rice varieties are diverse, the uneven stages of rice development provide abundant food for all generations, benefiting the occurrence of the pest. Improper management of fertilization and irrigation, causing the rice plants to be overly green and late-maturing, also facilitates damage by the pest (figure 9.10).

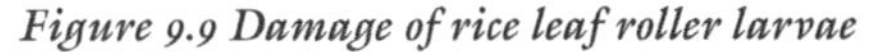

Figure 9.9 Damage of rice leaf roller larvae

Figure 9.10 Damage of rice leaf roller in the field

Comprehensive control of the rice leaf roller must adhere to integrated pest management strategies, combining healthy and high-yielding cultivation practices with coordinated use of biological and chemical control methods to keep the damage caused by the larvae within economically acceptable limits. Recommended approaches are as follows:

(1) *Promote healthy planting techniques.* Reform the cropping system, ensure a rational layout of seed varieties, and implement reasonable density planting to avoid mixing early, medium, and

late rice. Optimize rice cultivation techniques and promote simplified cultivation methods. Stick to scientific fertilization, enhance fertilizer and water management, apply sufficient base fertilizer in one go, and use nitrogen, phosphorus, and potassium reasonably. Topdressing fertilizer early and avoiding excessive nitrogen fertilizer can ensure that rice grows robustly and evenly. The rice will not grow excessively in the early stages and grow too long in the later stages. Then the rice can have better pest resistance and shorter harm period. Scientifically manage water and dry fields in a timely manner can reduce the field moisture during the larvae hatching period. In winter and early spring, remove weeds around the field when manuring can maintain field cleanliness and reduce the pest source base.

(2) *Select pest-resistant varieties.* On the premise of high-yielding cultivation, varieties with thick and hard leaves and tight main veins should be chosen. Super hybrid rice, having tall and strong stems, broad and hard leaves, and tight main veins, makes it difficult for young larvae of the rice leaf roller to roll the leaves. Thus, the pests have a lower survival rate, and generally causes less damage to super hybrid rice.

(3) *Implement biological control.* The natural enemies of the rice leaf roller include more than 60 species, which requires protection for producing rice. Based on the investigation of the types and quantities of natural enemies, coordinate the timing, type, and method of chemical control. If applying pesticides during the regular time harms the natural enemies significantly, consider applying the pesticide earlier or later; if the pest population has reached the control threshold but the parasitism rate of the natural enemies is high, pesticide application might not be necessary. When choosing the type and method of pesticide application, try to use pesticides and methods that do not harm or minimally harm natural enemies. Artificially release Trichogramma chilonis in the field, and before releasing, well know the occurrence period and population of the pest. From the beginning of the moth emergence period to the decline after the peak, release every two to three days, for two to three consecutive times. The amount of release depends on the field egg quantity of the rice leaf roller, for clusters with less than five eggs, release about 10,000 wasps per *mu* each time; for clusters with about ten eggs, release 30,000 to 50,000 wasps per *mu* each time. Bacillus thuringiensis is a microbial insecticide that has been used for a long time. Additionally, there are microbial pesticides like the Nucleopoly hedrovirus of the cabbage looper, Beauveria bassiana, Bacillus firmus, Bacillus sphaericus, Metarhizium anisopliae for caterpillars, and Metarhizium anisopliae for scarab beetles.

(4) *Implement chemical control.* Rice plants are particularly susceptible to damage from the rice leaf roller during the tillering and booting stages, especially the booting stage, where losses are greater. The optimal period for control is when the second and third instar larvae are at their peak population. Chemical control is appropriate just before the third instar larvae appear (i.e., when many leaf tips are rolled), especially in rice fields that are tender and green, which should be the focus of control efforts. Chemical control should adopt the strategy of "firmly controlling the second generation, cleverly controlling the third generation, and selectively controlling the fourth generation," with the possibility of treating the third and fourth-generation larvae in combination with other diseases and pests, depending on the situation. In production, pesticide application should be according to control thresholds, with the hybrid rice pest threshold being 50–60 per hundred tillers during the tillering stage and 30–40 per hundred tillers during the heading stage. Common pesticides and the application rate per *mu* include 20% chlorantraniliprole at 15 ml, 40% chlorantraniliprole · thiamethoxam at 15 g, or 5.7% emamectin benzoate at 80–100 g, mixed with 50 kg of water

for spraying. The use of plant protection UAVs for pesticide application is generally effective after the dew has dried in the late afternoon, evening, or morning, and can be applied on cloudy and light rainy days. Additionally, newer pesticides include tetrachlorantraniliprole, indoxacarb, abamectin · chlorfenapyr, and abamectin · indoxacarb. During the application period, shallow water of three to six cm should be irrigated and maintained for three to four days.

(5) *Use physical pest control methods.* Utilize insect pheromone technology to fully leverage sex attractants for the rice leaf roller, which can attract and kill a large number of male moths, leading to an imbalance in the sex ratio in the field, lowering the mating rate, and reducing the offspring pest population. Currently, the commonly used devices in production include white delta traps, cage traps, and water pan traps. These should be uniformly deployed over a large area, with one to two traps per *mu* of rice field for effective results. Additionally, use light-controlled frequency-vibrating insecticidal lamps and fan-suction type solar insecticidal lamps. Every three to four ha installed one insecticidal lamp. Lamps in rice fields can effectively reduce the basal population of rice leaf rollers.

3. Rice planthoppers

Rice planthoppers are commonly referred to as water rice mite in some Chinese contexts. The main types of planthoppers that pose a threat to rice crops are the brown planthopper and the white-backed planthopper. Among these, the brown planthopper is the most prevalent and causes the most severe damage. It is a migratory pest that prefers warm temperatures, and rice-growing regions south of the Yangtze River in China are almost annually affected by its infestations (figure 9.11).

Figure 9.11 Damage of cluster rice planthoppers

Figure 9.12 Damage of rice planthoppers in the field

In Hunan, the main pest affecting early-season rice is the white-backed planthopper. During the first crop and early stages of the second crop, a mixture of white-backed planthoppers and brown planthoppers occurs. However, in the later stages, the population of brown planthoppers rapidly increases. Adult and nymph-stage brown planthoppers prefer shady and humid environments, and they tend to gather at the base of rice plants. They feed on the sap of rice plants using piercing-sucking mouthparts and secrete toxic substances from their salivary glands, which leads to the withering of the rice plants. When the damage is mild, the lower leaves of rice plants turn yellow. When the damage is severe, the affected rice plant tissues gradually turn black and decay. When the infestation is severe, large sections of rice plants become withered, bending over, or collapsing completely, which is commonly referred to by local people as "burning," "piercing the top," or "yellow marsh." This ultimately leads to a significant reduction in yield or even complete crop failure (figure 9.12).

The brown planthopper exhibits five to seven generations per year in provinces such as Jiangsu, Zhejiang, Hunan, Jiangxi, etc. There is a noticeable phenomenon of overlapping generations in these areas. Adult brown planthoppers are classified into two types distinguished by their wing length: long-winged and short-winged. Long-winged adults possess migratory behavior and are attracted to light, making them migratory types. Short-winged adults, conversely, are non-migratory and have a strong reproductive capacity. They can complete a reproductive cycle in just over 20 days during the rice growing season. Under normal circumstances, each female can lay 200 to 500 eggs. The brown planthopper thrives and reproduces in temperatures ranging from 25°C to 28°C and relative humidity of 80% or higher. It is more likely to cause damage during rainy summers and mild autumns, rather than hot summers or cool late autumns. An increase in the population of short-winged adults in the fields, accompanied by a significant rise in reproduction, serves as a precursor to severe infestations. The white-backed planthopper, like the brown planthopper, is a migratory pest. It tends to migrate earlier than the brown planthopper. Its peak reproductive period coincides with the tillering and elongation stages of rice plants. The long-winged adults (males) have strong flying abilities, while the presence of short-winged adults is not seen. Female white-backed planthoppers have lower reproductive capacity compared to brown planthoppers, with an average of 85 eggs laid per female. The population distribution of white-backed planthoppers in the field is relatively even, and the multiplication rate of each generation is lower, resulting in generally milder damage compared to the brown planthopper. However, it is important to note that white-backed planthoppers can transmit the southern rice black-streaked dwarf virus, and therefore, proper pesticide control measures should be strengthened in rice seedbeds.

The comprehensive prevention and control of rice planthopper can be summarized as based on agricultural control, with resistant varieties as the main part, protecting natural enemies to reduce the population, utilizing ducks and frogs to eliminate pests, taking proactive measures to control and prevent, and using pesticides rationally to ensure a bumper harvest. The recommended approaches are as follows:

(1) *Adhere to agricultural prevention and control.* Under the conditions of mixed cropping of single- and double-cropping rice, nutrient-rich soil is favorable for the reproduction and damage of brown planthoppers. To prevent the proliferation and migratory damage of planthoppers, it is necessary to reform the unreasonable cropping system, implement contiguous planting, and adopt a rational layout of rice varieties (combinations). Enhancing cultural practices and nutrient management is essential. Pay attention to proper plant spacing and ensure adequate fertilization. Apply sufficient basal fertilizer and implement timely and targeted topdressing. Properly control nitrogen, increase potassium, and supplement phosphorus to avoid excessive nitrogen application and prevent excessive vegetative growth in the later stages of rice cultivation. Water management should include shallow water irrigation and frequent watering. The fields should be sunned at the appropriate time

to maintain ventilation, light transmission, and reduce field humidity. These measures can help reduce the reproductive rate of planthoppers. During the harvest of early-season rice, straw should be promptly removed and not piled up at the field edges. After harvest, irrigate and plow immediately. For late-season rice, it is recommended to transplant it early and apply spray treatments along the field edges for protection.

(2) *Promoting resistant varieties (combinations).* Promoting the use of resistant rice varieties (combinations) is the most economically effective measure for preventing and controlling brown planthoppers. Over the past 20 years, China has selected, evaluated, and utilized a large number of varieties (combinations) that are resistant to brown planthoppers. Additionally, there are varieties that have been developed through antigenic hybridization for insect resistance. In production, it is advisable to select and use insect-resistant varieties that are suitable for local cultivation. According to observation, brown planthoppers feeding on resistant varieties exhibit lower feeding quantities, slower development, higher mortality rates, and lower survival rates. Additionally, the occurrence of short-winged adult brown planthoppers in late-stage rice crops is also significantly reduced. Furthermore, on resistant varieties, the insect population density is low, making it difficult for them to form dominant populations. Currently, there are not many widely promoted super hybrid rice varieties with high resistance or multiple resistance combinations in production. However, the majority of these varieties possess moderate resistance or are situated between moderate resistance and moderate susceptibility. In particular, super hybrid rice plants are tall with sturdy stems, thick stem walls, and strong cuticles, which greatly enhance their tolerance to brown planthoppers.

(3) *Protecting and utilizing natural enemies.* Rice field spiders and black shoulder bugs are important predatory natural enemies of brown planthoppers. According to the investigation in Xiangyin County, Hunan Province, the dominant population of field spiders consists of nine species: Pardosa pseudoannulata, Lycosa sinensis, Pirata subpiraticus, Pardosa pseudoatrata, Hylyphantes graminicola, Pardosa youngi, Xerolycosa nemoralis, Pardosa astrigera, and Agyneta lobata. These nine species collectively account for 73.6% of the total spider population, with individual proportions of 15.70%, 12.80%, 10.70%, 9.37%, 7.44%, 7.16%, 4.13%, 3.30%, and 3.00% respectively. The other dominant populations of predatory natural enemies include black shoulder bugs (accounting for 39.5%), black water skater (accounting for 18.42%), and green lacewings (accounting for 11.80%). These three species collectively make up 69.72% of the total population of such natural enemies. Parasitic natural enemies include rice planthopper braconids, rice planthopper nematodes, and parasitic wasps. In addition, there are also ladybugs, ground beetles, tiger beetles, predatory bugs, and assassin bugs as natural enemies. It is important to protect these natural enemies and fully utilize their control effect on brown planthoppers. During the spring planting and double-cropping periods, it is recommended to flood and plow the fields before placing rice straw to assist in the migration or to create protective pits. It is also beneficial to promote the planting of beans on field ridges and maintain green corridors by protecting field ridges and roadside weeds. Additionally, human intervention can help ensure the safe migration of populations of natural enemies such as spiders. It is advisable to use highly efficient and low-toxicity pesticides, as well as biopesticides, to minimize harm to natural enemies.

(4) *Raising ducks in rice fields to prevent and control pests.* Due to their strong foraging ability, gregarious nature, and affinity for water, ducks are suitable for free ranging in rice fields. It is recommended to select robust duck breeds with wide adaptability, high egg production, and medium-sized body. Some excellent duck breeds for this purpose include the "Jiangnan No. 1" duck, the Sichuan

Ma duck, and the Linwu duck, among others. Typically, 15 ducklings or 10 adult ducks are allowed to roam freely per *mu* of rice field. The "wide-narrow row" planting method is implemented to provide convenience for ducks to move around and forage. Plastic mesh fences can be set up along the field embankments to prevent the ducks from escaping. The ducks are released into the rice fields after 18 to 20 days from when the ducklings hatch or 15 to 20 days after rice transplantation. Adult ducks can be directly introduced into the rice fields to feed on pests such as rice planthoppers. However, it is important to note that during the milk ripening to maturing stage of the rice plants, ducks should be kept out of the fields to prevent them from eating the rice grains.

(5) *Implement chemical control measures.* Based on the type of rice variety and the occurrence of rice planthoppers in the field, chemical control should be carried out timely. Inspect the developmental progress and determine the appropriate timing for pest control. The peak periods for chemical control against brown planthoppers and white-backed planthoppers are typically during the second and third instar stages of their nymphs. Assess the density of insect mouths and identify the fields that require control measures. After the peak hatching period of eggs, investigate whether the density of various rice field pests has reached the control threshold.

Any rice field that meets the control threshold should be designated as a field for pest control. The control threshold is determined based on the density of rice planthoppers, with the following guidelines: for conventional rice varieties, the threshold is 1,000 to 1,500 insects per hundred clusters during the heading stage; for hybrid rice varieties, the threshold is 1,500 to 2,000 insects per hundred clusters during the heading stage; for super hybrid rice varieties, the threshold is 2,500 to 3,000 insects per hundred clusters during the heading stage. The "control before the peak, pressure after the peak" strategy is adopted for the control of rice planthoppers in mid- and late-season rice. High-efficiency, low-toxicity, and long-lasting pesticides are selected. Examples of recommended pesticides and their application rates are as follows: 16ml/*mu* of 10% triflumezopyrim, 20–30 g/*mu* of 80% dienolpyridoxone, 20 ml/*mu* of 60% pymetrozine, 40–50 g/*mu* of 25% buprofezin wettable powder, 20–30 g/*mu* of 10% imidacloprid wettable powder, 50–60 g/*mu* of 25% pymetrozine wettable powder, or 5–10 g/*mu* of 25% thiamethoxam water dispersible granules. These pesticides are highly effective against rice planthoppers and have a residual effect lasting for 20–30 days. Generally, a single application can control the pest throughout the entire growth period. However, for super hybrid rice with a longer growth period, two applications may be required, and the timing of pesticide application should be controlled at the peak period of nymphs in young age. In addition to the mentioned commonly used pesticides, there are also other options available: 30–40 ml/*mu* of 10% Nitenpyram water-based formulation, 10–15 g/*mu* of 40% chlorantraniliprole + thiamethoxam (Fipronil), 10–15 ml/*mu* of dinotefuran; 10–15 ml/*mu* of 20% mefenacet soluble liquid formulation, or 5–10 g/*mu* of 70% confidor water dispersible granules. When brown planthoppers reach epidemic levels or re-infestation occurs (with more than 3,000 insects per hundred clusters), additional quick-acting pesticides such as 80% dichlorvos or tsumacide or 150–200 ml of isoprocarb emulsion can be used during the late stage of rice growth to rapidly control the pests. Choose one of the mentioned pesticides and mix it with 50–60 kg of water per *mu* for manual spraying. Pay attention to rotate and alternate the use of pesticides. Ensure that the spray is evenly distributed at the base of the rice plants. Alternatively, UAVs can be used for ultra-low-volume spraying.

4. Rice thrips

The common name for rice thrips is "gray planthopper." It is a significant pest in the early stages of rice growth, particularly during the seedling and tillering stages in the main crop field (figure 9.13, figure 9.14).

Rice thrips infestation is more severe in super hybrid rice and late-season rice seedlings, resulting in greater damage. The adult and nymphs both use rasping-sucking mouthparts to scrape and suck plant sap from the leaf surface. In the early stages of infestation, small white to yellow-brown spots appear on the leaves. Subsequently, the leaf tips curl and wither due to water loss, eventually leading to complete leaf curling and yellowing. The adult planthopper is black-brown in color and has a small body, resembling an ant. The nymphs, on the other hand, are pale white to yellow-brown in color and have an even smaller body size. They are light-sensitive and often gather in large numbers on the affected leaves. In severe cases, the infestation can cause widespread wilting and curling of the rice seedlings, resulting in loss of green coloration throughout the paddy field. The rice thrips has a short life cycle, fast reproduction rate, and multiple generations within a year in the Yangtze River region. Approximately 10 to 14 generations occur annually. The adult insects overwinter on grasses such as wheat, wild grass, or Alopecurus aequalis (a type of weed). After the rice seedlings have emerged, a large number of adult planthoppers migrate into the nursery fields and later move from the nursery fields to the main crop fields, causing damage. They exhibit a preference for young and green plants and tend to gather in groups. The optimal temperature range for the growth and reproduction of the rice thrips is around 20°C to 25°C. It requires a relative humidity of over 80%. It is tolerant to low temperatures but not to high temperatures. When the temperature rises above 28°C, its survival is hampered, and the population size decreases. There are several sensitive periods during which rice is vulnerable to damage from rice thrips: the second to fifth leaf stage of seedlings, the tillering stage, and the early panicle differentiation stage. It is important to enhance preventive measures during these periods. The recommended approaches are as follows:

Figure 9.13 Symptoms of damage by rice thrip in the seedbed

Figure 9.14 Symptoms of damage by rice thrip on individual leaves

(1) *Clearing weeds along the field edges to reduce the population of insects.* Combining this with winter fertilization and clearing weeds both inside and outside the nursery fields can help reduce overwintering sources and early spring breeding of intermediate hosts, preventing the migration and subsequent damage caused by rice thrips.

(2) *Rational variety layout and continuous planting of rice fields are recommended.* For both single- and double-cropping rice, as well as for the same variety, it is encouraged to plant them continuously in a contiguous manner, avoiding intercropping and minimizing mixed planting. This helps to prevent the worsening of feeding conditions for rice thrips. Ensuring consistent rice growth is important. In terms of variety types, hybrid rice is more susceptible to damage than conventional rice, and super hybrid rice is more susceptible to damage compared to general hybrid rice.

(3) *Foster strong seedlings and implement scientific fertilization practices.* Emphasis should be placed on strengthening the management of nursery fields for mid-season and late-season rice, with the goal of cultivating robust seedlings. For nursery fields that have already been affected by rice thrips, an additional application of quick-acting fertilizer should be applied after pesticide treatment to promote the recovery of seedling growth.

(4) *Protect natural enemies for biological pest control.* Key predatory natural enemies of adult and nymph stages of rice thrips include spiders (such as micro-spiders), Verania discolor, Orius minutus, and yellow brown flower bugs. And parasitic natural enemies are mainly nematodes, etc. These natural enemies have a certain inhibitory effect on the occurrence of rice thrips.

(5) *Master the appropriate timing and apply pesticides in a timely manner.* Timing for pesticide application should be based on seedling and pest conditions, with a focus on preventing and controlling brown planthoppers in mid-season and late-season nursery fields, as well as during the tillering stage of the crop.

The recommended threshold for pesticide treatment is when the leaf rolling rate reaches 10%–15% in nursery fields, or when there are 100–200 planthoppers per hundred plants. In the main rice field during the tillering stage, treatment is advised when the leaf rolling rate reaches 20%–30%, or when there are 200–300 planthoppers per hundred plants. It is important to conduct timely pesticide treatment under these circumstances. When selecting pesticides, it is preferable to choose highly effective, low-toxicity, and long-lasting pesticides. For example, use 20 g/*mu* of 10% imidacloprid wettable powder, or 20 g/*mu* of 25% pymetrozine wettable powder, or 30 g/*mu* of 25% buprofezin wettable powder. These pesticides can be mixed with 150 g of urea per *mu* in the spray solution, which can not only control the pests but also promotes the growth of seedlings. Additionally, if the pest population is high and causing significant damage, and most leaves are curled and yellowed, quick-acting pesticides can be used. For instance, using 100 ml/*mu* of 20% triazophos emulsion, or 1,500-fold dilution of 90% metrifonate crystal, or 200 ml/*mu* of 25% insecticide water agent. The recommended pesticide dosages per *mu* are based on mixing with 50 kg of water for spraying. Furthermore, during sowing, mixing 10–20 g of 70% imidacloprid wettable powder with every 10 kg of paddy dry seeds can effectively control rice thrips and rice planthopper in the nursery fields for a period of over 30 days, yielding excellent results.

III. Weed Control Techniques

1. Different cultivation types in rice fields in Hunan Province

The weed species in rice fields in China are diverse, with high density, wide coverage, and significant harm. The application of chemical weed control techniques in rice fields is an important aspect of achieving modern agriculture. Investigations show that there are more than 60 common weeds in rice fields, including major weeds such as barnyardgrass, Potamogeton distinctus, and pickerelweed, which are among the top ten weed pests in China. Barnyardgrass, in particular, poses a serious threat, with large densities observed in many fields, overpowering the conventional rice plants. In some hybrid rice seedling fields, the phenomenon of "three-story" growth has also been observed due to barnyardgrass infestation. Therefore, barnyardgrass is one of the most severe weeds in rice fields (figures 9.15 and 9.16).

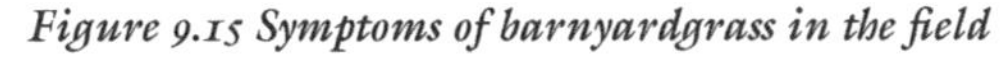

Figure 9.15 Symptoms of barnyardgrass in the field

Figure 9.16 Symptoms of pickerelweed in the field

Eclipta alba (千金子), is an important weed in direct-seeded rice fields in the middle and lower reaches of the Yangtze River region in China. Its harmfulness is second only to barnyardgrass. Other significant weeds in these rice fields include Eleocharis yokoscensis (牛毛毡), Cyperus difformis (异型莎草), Sagittaria pygmaea (矮慈姑), Equisetum ramosissimum (节节草), Alternanthera philoxeroides (空心莲子草), and Marsilea quadrifolia (四叶萍). The occurrence and distribution of weeds in different regions of Hunan Province are generally similar, but there may be some variations due to differences in topography, soil conditions, and cultivation methods. The extent of weed damage is closely related to cultivation practices such as direct-seeding, transplanting, and water management. For instance, weed infestation in broadcast-seeded fields is often more severe than in transplanted fields.

In the past 20 years, there have been significant changes in rice cultivation methods in China. In the middle and lower reaches of the Yangtze River, the traditional practice of manual transplanting has been replaced by manual direct-seeding, mechanical direct-seeding, mechanized seedling using plastic trays or bowls—mechanical transplanting, and manual seedling scattering using plastic trays—manual broadcasting seedlings. However, some regions still maintain the traditional method of manual transplanting. Due to the differences in planting and water management methods, the density and severity of weed populations vary. Therefore, in production, it is necessary to implement scientific chemical weed control based on the population and characteristics of weeds under different rice cultivation conditions (such as planting and water management). Different types of rice field weeds require specific control techniques, including the selection and application methods of herbicides. In order to carry out chemical weed control in rice fields in a scientific and reasonable manner, weed experts in Hunan Province have conducted extensive investigations and research over the years, combined with practical production. They have categorized rice production in Hunan into nine different types, ranging from seedling cultivation to field transplanting. According to different seedling cultivation methods, planting methods, and herbicide application techniques, weed control technical protocols have been developed for each of these nine cultivation management types in rice fields. These protocols have wide applicability, strong operability, mature technology, and convenient use, and they have shown good results in weed control.

2. Main weed species in rice fields in Hunan

In rice fields across the country, weeds are classified into three categories. The first category includes grassy weeds from the Poaceae family, such as barnyardgrass, Eclipta alba, and barnyard millet. The second category consists of broadleaf weeds like Sagittaria trifolia, Ludwigia octovalvis, and pickerelweed. The third category includes sedge weeds from the Cyperaceae family, such as Cyperus difformis, Fimbristylis miliacea, and Eleocharis kuroguwai.

The weed species are generally similar in early water-seeded fields, late water-seeded fields, transplanted fields, direct-seeded with seedling broadcast, and direct-seeded with plowing. The main weed species in these fields are the first 13 mentioned above. Dry-seeded fields and non-tillage direct-seeded fields mainly have the latter 7 species, along with barnyardgrass, Semen Euphorbiae, Aeschynomene indica, pickerelweed, and Cyperus difformis.

3. Weed control techniques for different types of rice fields

In production, it is important to choose different herbicides (formulations) and application methods based on the weed species and their characteristics in different types of rice fields. Herbicide selection and usage can be categorized into chemical weed control in nursery fields, chemical weed control in main fields (transplanting, seed broadcasting, machine transplanting), and chemical weed control in direct-seeded fields. When considering the weed species and weed community composition in the field, the selection of herbicides should prioritize the target weeds, spectrum of efficacy, and control effectiveness, while also considering the specific characteristics and cost of the herbicides. In rice fields in Hunan, the herbicide butachlor is widely used, primarily for the control of barnyardgrass, Eleocharis yokoscensis, and ricefield flatsedge. Additionally, to achieve comprehensive control, it is necessary to mix or combine different herbicide varieties, such as pendimethalin and bensulfuron-methyl.

Chemical herbicides are classified as toxic pesticides, so it is important to select herbicides that are highly effective, have a broad spectrum, are low in toxicity, and are safe for humans, livestock, and the ecological environment. It is crucial to strictly adhere to the Standards for Safe Use of Pesticides and follow the instructions provided in the product labels when applying herbicides. Based on the actual occurrences of weeds in rice fields in Hunan and the techniques of chemical weed control, weed control experts have conducted comprehensive analysis and technical summaries. They have developed technical regulations for weed control in different types of rice fields, taking into account the specific cultivation characteristics. It is advisable for each region to choose the appropriate weed control techniques based on their actual conditions. Here are nine different types of rice field weed control techniques, as researched and summarized by herbicide experts in Hunan Province, China.

1) *Weed control techniques in early rice hydroponic (moist) seedling fields.* Early rice water seedling cultivation is also known as seedling cultivation in a moist environment, which is a conventional method to cultivate young or medium-sized rice seedlings.

   a) For the moist nursery of early rice, the field surface should be leveled and kept moist without any stagnant water. A mud slurry should be applied through trampling before sowing. After four days of seeding, during the stage of seedling emergence, spray 80–100 ml of 30% pretilachlor + safener EC mixed with 30–45 kg of water, per *mu*.
   b) During the stage of seedling emergence, three to four days after seeding, spray 50–60 g of 40% pretilachlor + safener + bensulfuron-methyl soluble powder, mixed with 30–45 kg of water, per *mu*.
   c) When the seedlings have grown to the two-to-three-leaf stage, spray 40–50 ml of 2.5% fluazifop-P-butyl oil suspension, mixed with 20–30 kg of water, per *mu*. Either option can effectively control barnyardgrass, annual sedges, and broadleaf weeds.

2) *Weed control techniques for wet nursery fields in late-season rice (or single-cropping rice).* Wet nursery is the most commonly used method for cultivating late-season rice and single-cropping rice seedlings. It helps to maintain soil moisture and aeration, which is beneficial for the growth of seedling roots.

   a) For wet nursery fields in late-season rice or single-cropping rice, it is also necessary to ensure a smooth surface and maintain moisture without waterlogging. Mud slurry should be used during seeding. During the seedling emergence stage, spray 80–100 ml of 30% pretilachor (propaquizafop + safener) emulsion, mixed with 30–45 kg of water, per *mu*.
   b) During the stage of seedling emergence, three to four days after seeding, spray 50–60 g of 40% propaquizafop + safener + bensulfuron-methyl soluble powder, mixed with 30–45 kg of water, per *mu*.

c) When the seedlings have grown to the two-to-three-leaf stage, spray 40–50 ml of 2.5% fluazifop-P-butyl oil suspension, mixed with 20–30 kg of water, per *mu*. Any of these three methods can effectively control barnyardgrass, annual sedges, and broadleaf weeds.

3) *Weed control techniques for dry nursery fields.* Typically, dry nursery fields are established on upland or vegetable garden soil beds or rice fields. In dry nursery fields, both aquatic and terrestrial weeds can coexist and occur in large quantities. When selecting herbicides, it is important to consider controlling both types of weeds simultaneously.

a) After sowing, water the field until the runoff drains naturally. Once the water has drained, spray 100 ml of 40% pendimethalin · ortho herbicide emulsion, mixed with 30 kg of water per *mu*. Then, during the seedlings' three-leaf stage, spray evenly 40–50 ml of 2.5% fluazifop-P-butyl oil suspension combined with 40 ml of 10% quizalofop-P-tefuryl emulsion, mixed with 20–30 kg of water.
b) After sowing, water the field until the runoff drains naturally and the surface has dried. Then, spray 80 ml of 60% pendimethalin emulsion combined with 20 g of 10% bensulfuron-methyl soluble powder, mixed with 30 kg of water, per *mu*. During the seedlings' three-leaf stage, spray evenly 40 g of 50% dichloroquinoline acid-soluble powder, mixed with 30 kg of water.
c) After sowing, water the field until the runoff drains naturally. Once the water has drained, spray 50–60 g of 40% propaquizafop + bensulfuron-methyl soluble powder, mixed with 30 kg of water, per *mu*. Then, during the seedling's three-leaf stage, spray evenly 40–50 ml of 2.5% fluazifop-P-butyl oil suspension, or 60 ml of 10% cyhalofop-butyl emulsion, mixed with 30 kg of water. Any of these three methods can effectively control crabgrass, yellow nutsedge, sedge grass, broadleaf weeds, as well as aquatic and terrestrial weeds.

4) *Weed control techniques in transplanted rice fields.* The key focus of weed control in transplanted rice fields is the first peak of weed emergence, and it is important to use herbicides in a timely and correct manner.

a) Five to seven days after transplanting early-season rice, or three to five days after transplanting mid- and late-season rice, apply 30 g of 20% butachlor · benfuresate · pretilachor wettable powder per *mu*, or 150 g of 30% butachlor · benfuresate · pretilachor wettable powder per *mu*. Mix them evenly with fertilizer and spread them or mix them with about 10 kg of fine sand for spreading. Maintain a shallow water layer of three to five cm in the field for about seven days after application.
b) Ten to fifteen days after transplanting early rice, or eight to twelve days after transplanting medium and late rice, spray 50 g of 50% dichlormid acid wettable powder and 30 g of 10% bensulfuron-methylwettable powder per *mu*, diluted in 30 kg of water. Drain the water from the field before spraying and irrigate the field with water within 24–48 hours after application. Maintain a shallow water layer of three to five cm in the field for about seven days.

c) Three to five days before rice transplanting, apply 150 g of 30% butachlor · benfuresate ·pretilachor wettable powder per *mu*, or 100 ml of 25% halosulfuron-methyl emulsion mixed with about 10 kg of fine sand, evenly spread in the water layer of the field. Fifteen to twenty days after transplanting, spray 60–80 ml of 2.5% fluorosulfuron oil suspension evenly. Drain the water from the field before spraying and irrigate the field with water within 24–48 hours after application. Maintain a shallow water layer of three to five cm in the field for about seven days.

5) *Weed control techniques in direct-seeded rice fields.* Weed control in direct-seeded rice fields requires high safety standards when selecting herbicides. Extra caution should be exercised in choosing the appropriate herbicides.

a) Five to eight days after direct-seeding rice, apply 150 g of 30% butachlor · benfuresate · pretilachor wettable powder per *mu*, or 60 ml of 50% pretilachor emulsion mixed with 30 g of 10% bensulfuron-methylwettable powder per *mu*. Mix them with fertilizer and spread them or mix them with about 10 kg of fine sand for spreading. Maintain a water layer of three to five cm in the field for about seven days after application.
b) Ten to fifteen days after direct-seeding rice, spray 50 g of 50% dichlormid acid wettable powder and 30 g of 10% bensulfuron-methylwettable powder per *mu*, diluted in 30 kg of water. Drain the water from the field before spraying and irrigate the field with water within 24–48 hours after application. Maintain a water layer of three to five cm in the field for about seven days after application.
c) Fifteen to twenty days after direct-seeding rice, spray 60–80 ml of 2.5% fluorosulfuron oil suspension per *mu*, diluted in 30 kg of water. Drain the water from the field before spraying and irrigate the field with water within 24–48 hours after application. Maintain a water layer of three to five cm in the field for about seven days after application.

6) *Weed control techniques in mechanized transplanted rice fields.* Mechanized transplanted rice fields employ the method of factory-produced seedlings, where specialized equipment is used to cover the seedlings with soil, sow the seeds, water them, and apply controlled electric heating to promote rapid germination and seedling emergence. This method allows for large-scale production of rice seedlings suitable for mechanized planting.

a) Six to eight days after machine-transplanting rice seedlings, apply 150 g of 30% butachlor ·benfuresate · pretilachor wettable powder per *mu*, or 60 ml of 50% pretilachor emulsion mixed with 30 g of 10% bensulfuron-methylwettable powder per *mu*. Mix them with fertilizer and spread them or mix them with about 10 kg of fine sand for spreading. Maintain a water layer of three to five cm in the field for about seven days after application.
b) Ten to fifteen days after machine-transplanting rice seedlings, spray 50 g of 50% dichlormid acid wettable powder and 30 g of 10% bensulfuron-methylwettable powder per *mu*, diluted in 30 kg of water. Drain the water from the field before spraying and irrigate the

field with water within 24–48 hours after application. Maintain a water layer of three to five cm in the field for about seven days after application.

c) Three to five days before machine-transplanting the seedlings, apply 150 g of 30% butachlor · benfuresate · pretilachor wettable powder per *mu*, or 100 ml of 25% halosulfuron-methyl emulsion mixed with about 10 kg of fine sand, evenly spread in the water layer of the field. Fifteen to twenty days after machine-transplanting, spray 60–80 ml of 2.5% fluorosulfuron oil suspension evenly. Drain the water from the field before spraying and irrigate the field with water within 24–48 hours after application. Maintain a water layer of three to five cm in the field for about seven days after application.

7) *Weed control techniques in plowed and broadcasted rice fields.* In fields with sufficient water supply and convenient irrigation and drainage systems, soil tillage, fertilization, and leveling are typically done before sowing rice seeds in shallow furrows. However, excessive water accumulation on the field surface should be avoided.

a) Three to five days after rice sowing, spray 100–120 ml of 30% (pretilachor + safener) emulsion per *mu*, or 40% (pretilachor + safener + bensulfuron), or 60% (butachlor + safener) emulsion per *mu*, diluted in 30 kg of water. Then, when the seedlings reach the three-to-four-leaf stage, spray 60 ml of 2.5% fluorosulfuron oil suspension, diluted in 30 kg of water. Irrigate the field within 24–48 hours after herbicide application and maintain a water layer of three to five cm in the field for about seven days.

b) Three to five days after rice sowing, spray 100–120 ml of 30% (pretilachor + safener) emulsion per *mu*, or 40% (pretilachor + safener + benfuresate), or 60% (butachlor + safener) emulsion per *mu*, diluted in 30 kg of water. When the seedlings reach the three-to-four-leaf stage, spray 50 g of 50% dichlormid acid wettable powder, diluted in 30 kg of water. Irrigate the field within 24–48 hours after herbicide application and maintain a water layer of three to five cm in the field for about seven days.

c) Three to five days after rice sowing, use the same herbicides and dosage as in the previous two methods. When the seedlings reach the three-to-four-leaf stage, spray approximately 50 ml of 2.5% fluorosulfuron oil suspension + 10% cyhalofop-butyl emulsion. Irrigate the field within 24–48 hours after herbicide application and maintain a water layer of three to five cm in the field for about seven days.

8) *Weed control techniques in non-tillage direct-seeded rice fields (rice-rapeseed rotation).* In the case of rice-rapeseed rotation, where rapeseed is the preceding crop and rice is directly seeded after rapeseed harvest, the coexistence period between weeds and rice is relatively long. It is important to choose effective herbicides.

a) When the field is moist but not waterlogged after sowing, spray 150–200 ml of 60% butachlor emulsion, or 100 ml of 50% pretilachor emulsion, or 125–150 ml of 33% clomazone mixed with 20–30 g of 10% benfuresate, diluted in 30 kg of water. When the seedlings reach the three-to-four-leaf stage, spray 100–120 ml of 2.5% fluorosulfuron oil suspension, or 100–150 ml of 10% cyhalofop-butyl emulsion, diluted in 30 kg of water.

Irrigate the field within 24–48 hours after herbicide application and maintain a water layer of three to five cm in the field for about seven days.

b) When the field is moist but not waterlogged after sowing, spray 150–200 ml of 42% butachlor · penoxsulam emulsion, diluted in 30 kg of water. When the seedlings reach the three-to-four-leaf stage, spray 100–120 ml of 2.5% fluorosulfuron oil suspension, or 100–150 ml of 10% cyhalofop-butyl emulsion, diluted in 30 kg of water. Irrigate the field within 24–48 hours after herbicide application and maintain a water layer of three to five cm in the field for about seven days.

c) When the field is moist, especially if there is waterlogging after sowing, spray 100 g of 40% pretilachor · bensulfuron-methyl (safener) wettable powder, diluted in 30 kg of water. When the seedlings reach the three-to-four-leaf stage, spray 100–120 ml of 2.5% fluorosulfuron oil suspension, or 100–150 ml of 10% cyhalofop-butyl emulsion, diluted in 30 kg of water. Irrigate the field within 24–48 hours after herbicide application and maintain a water layer of three to five cm in the field for about seven days.

9) *Weed control techniques in non-tillage direct-seeded rice fields (fallow rice fields).* In the case of leisure rice fields during the winter and spring seasons, similar to the previous scenario with rapeseed preceding the rice crop, direct-seeding without tillage allows weeds to establish a large population. Additionally, some of the weeds may be older, which can significantly affect the growth of rice seedlings. Therefore, it is important to implement effective weed control measures.

a) Three to five days after sowing, spray 60 g of 40% pretilachor · bensulfuron-methyl (safener) wettable powder, diluted in 30 kg of water, per *mu*. When the seedlings reach the three-to-four-leaf stage, spray 60–80 ml of 2.5% penoxsulam oil suspension, diluted in 30 kg of water. Irrigate the field within 24–48 hours after herbicide application and maintain a water layer of three to five cm in the field for about seven days.

b) At the two-leaf-one-bud stage of the seedlings, spray 50 g of 50% quinclorac acid wettable powder mixed with 30 g of 10% bensulfuron, diluted in 30 kg of water, per *mu*. When the seedlings reach the three-to-four-leaf stage, spray 80–100 ml of 10% cyhalofop-butyl emulsion, diluted in 30 kg of water. Irrigate the field within 24–48 hours after herbicide application and maintain a water layer of three to five cm in the field for about seven days.

c) Three to five days after sowing, spray 100 ml of 60% butachlor (safener) emulsion, diluted in 30 kg of water, per *mu*. Then, when the seedlings reach the three-to-four-leaf stage, spray around 50 ml of 2.5% fluorosulfuron oil suspension + 100 ml of 10% cyhalofop-butyl emulsion, diluted in 30 kg of water. Irrigate the field 24–48 days after herbicide application and maintain a water layer of three to five cm in the field for about seven days.

4. Precautions for using different herbicides in rice fields

1) *In seedling fields of early-season rice, late-season rice, and early seedlings, if there are a few tall weeds such as barnyard grass, grassy weeds, or sedge grass, they should be promptly removed to avoid carrying them into the main field during transplanting.*
2) *It is generally not recommended to use herbicides containing acetochlor in small and weak seedling fields, as well as fields with leaking water.* For larger and stronger seedlings and fields with good water retention, you can use herbicides like 20% acetochlor · bensulfuron-methyl or 30% butachlor · bensulfuron-methyl and other herbicides. However, after herbicide application, if heavy rainfall occurs, drainage should be ensured promptly to protect the seedlings. Closed weeding control can be used for two to three times before and after transplanting, and if carriers such as sand, soil, or fertilizer are used, they must be evenly mixed and applied.
3) *Herbicides containing acetochlor should not be used in fields where rice is broadcasted, or machine transplanted.* Instead, herbicides like 30% butachlor · bensulfuron-methyl or 40% pretilachor · bensulfuron-methyl can be used, along with other appropriate herbicides. However, if heavy or torrential rain occurs, timely drainage should be carried out to protect the seedlings.
4) *After applying soil-incorporated herbicides in fields prepared through tillage, ensure that the water level in the field does not submerge the heart leaves of the seedlings.* If it rains, drainage should be carried out promptly. When applying foliar-applied herbicides, spray them as a fine mist to facilitate absorption and translocation within the stems and leaves, thereby improving herbicidal efficacy.
5) *When using soil-incorporated herbicides in non-tillage direct-seeding fields of oil rice, attention should be paid to ensure that the rice seeds are adequately covered with soil.* If the rice seeds are exposed, the standing water in the field may cause phytotoxicity. If heavy or torrential rain occurs after herbicide application, timely drainage should be carried out to protect the seedlings.
6) *In non-tillage direct-seeded fallow fields, if using 50% dichlormid acid, it should be done during the seedling stage of rice, and its use should be stopped after the rice reaches the tillering stage to avoid herbicide injury.*
7) *Regardless of transplanting or direct-seeding, if perennial sedge grass (such as Sparganii Weed) appears and the harm is serious, herbicides such as bentazone or carfentrazone plus MCPA can be used for control.* It is noted that the latter should be used at the late tillering stage of rice. Additionally, in direct-seeded fields, intermittent older barnyardgrass and Eclipta alba, etc., can be manually removed during topdressing.
8) *Herbicide applications should be conducted on sunny days.* Before application, carefully inspect the pesticide application equipment to prevent the leakage of the drug. Choose a safe location for dispensing. During field operations, wear masks, gloves, and long-sleeved clothing, taking personal protective measures. Pay attention to the wind direction. After herbicide application, wash hands and clean the equipment thoroughly.

IV. Crop Protection UAV Application Techniques

1. The advantages and characteristics of Crop Protection UAVs

China is a large agricultural country, and the pests and diseases of crops such as rice are rampant. The use of pesticides still plays a significant role in pest control. Due to the long-term constraints of outdated pesticide spraying equipment and its use technology, traditional methods of pesticide application are still prevalent in most areas. These methods include manual backpack sprayers, handheld spray guns, and various motorized sprayers. These spraying devices require a long operation time, involve high labor intensity, require large amounts of pesticides, and result in high costs. As a result, the pesticide utilization rate has remained low, with only 20%–30% of the pesticides depositing on the target area, while the rest leads to pollution and wastage. In the past decade, China's agriculture has undergone a transition from traditional to modernization, with developments and innovations in the aspects of production methods, technologies, systems, and mechanisms. To strengthen and standardize specialized pest control services for crops and enhance the capacity for socialized plant protection services, crop protection UAVs have received attention in agricultural production. The application of UAV technology in pesticide spraying has gradually gained momentum, especially in the control of pests and diseases in rice fields. UAVs have become a new type of plant protection machinery in agricultural production, offering advantages such as easy operation, flexible use, labor and cost savings, fast and efficient operation, and uniform spraying. As a result, they have gained widespread acceptance and support from farmers. Crop protection UAVs operate at a height of two to three m above the crops, resulting in minimal drift. The downward airflow generated by the rotor assists in improving the penetration of the pesticide fog onto the crops, enhancing its efficacy. This technology can save pesticide usage by 10% to 20% and cover an area of three to four ha per hour for pest control. With the increase in cooperative and contracted farming households, agricultural production has shifted toward contiguous cultivation and large-scale operations. It has become challenging to rely solely on manual labor for crop protection in large-scale agricultural production. Moreover, the cost of employing manual labor for pesticide spraying has been rising. This has led to an increasing demand for mechanized plant protection equipment in agricultural production. In comparison to traditional methods of pesticide application, agricultural UAVs demonstrate significant advantages and characteristics in plant protection operations.

2. Structural design principles of crop protection UAVs

According to current agricultural practices, crop protection UAVs can be classified into three types based on their power source: electric, gasoline, and hybrid. Electric UAVs typically use lithium batteries, offering a simple and lightweight design. They are flexible, easy to maintain, and adaptable to various conditions. However, they may have limited wind resistance and endurance capabilities.

Gasoline-powered UAVs, on the other hand, utilize fuel as their power source. They can carry heavier payload, possess strong wind resistance, and have longer flight endurance. However, they may require higher operational skills for autonomous flight. Hybrid UAVs combine both electric and gasoline power sources, aiming to leverage the advantages of both types while compensating for their respective limitations. However, there are still some technical challenges to overcome. Crop protection UAVs can also be classified based on their structural design into three types: fixed-wing UAVs, single-rotor UAVs, and multi-rotor UAVs. Fixed-wing UAVs have a larger payload capacity, higher flight speeds, and greater operational efficiency. However, they require suitable takeoff and landing areas without obstacles. Single-rotor UAVs offer stable flight in windy conditions and provide good atomization effects. They generate strong downward airflow and have the ability to penetrate vegetation, allowing the spray to reach the base of crop stems. However, they are expensive, difficult to operate, have poor safety, and are prone to malfunction. Multi-rotor UAVs have a moderate price and are easy to operate, making them suitable for small fields. However, they have relatively weak wind resistance and limited coverage and spraying range compared to other types. Crop protection UAVs mainly consist of power systems, spraying systems, and flight control systems. The power systems can be categorized into electric, gasoline, and hybrid types, among which the electric power system includes components such as power batteries, motors, electronic speed controllers, and rotors, providing the main power source for the UAV's flight. The spraying system is the key component responsible for pesticide application over crops. It allows flexible and precise spraying operations as per the requirements. The flight control system is the human-controlled system of the UAV. It provides control and decision-making commands, performs functions like GPS positioning, navigation, and data collection, and can be operated remotely for completing the tasks. Currently, there are numerous brands of crop protection UAVs with varying performance and significant differences in technological capabilities. Additionally, the disparity in operator skills often results in differences in operational quality. When selecting a crop protection UAV, certain factors should be considered. First, it is important to determine whether the UAV meets the requirements outlined in the Quality Evaluation Technical Specification for Crop Protection UAVs issued by the Ministry of Agriculture and Rural Affairs of China (NY/T 3213—2018). Second, priority should be given to well-established brands in the industry, such as DJI, EHang, Hanhe, and Yuren. These brands generally have higher technological capabilities and are known for their reliability and adaptability, placing them at the leading position in the industry. Additionally, considering the actual operational skills of UAV operators is crucial. Providing training to the technical team responsible for operating the crop protection UAVs is essential for improving their skills, which ultimately impacts the quality of operations.

3. The key technologies of crop protection UAVs

(1) *Monitoring and reporting technology.* The development and application of crop protection UAVs in agricultural production belong to new technological products. Currently, these UAVs are mainly used by professional crop protection companies. As a professional crop protection company, it must first be equipped with highly qualified professionals who can formulate a year-round crop protection

program specifically for major diseases and pests affecting crops such as rice. These plans need to provide detailed rice disease and pest control strategies for large-scale farmers or households and sign a control agreement based on these strategies. Especially before each application of pesticides, it is essential to have professionals conduct field monitoring of diseases and pests and ensure accurate reporting. It becomes possible to accurately predict the timing and severity of disease and pest occurrences by monitoring the occurrence of diseases and pests during different growth stages of rice, understanding the patterns of disease outbreaks, tracking pest development progress, and determining pest density and level of damage. Combining this information with practical experience, suitable timing for control measures and pesticide application can be determined. In general, it is important to apply pesticides during the early stages of disease and pest outbreaks, taking advantage of favorable periods, and utilize crop protection UAVs for timely spraying. Accurate monitoring and timely application of pesticides are crucial. For example, insecticides should be applied during the peak hatching period of insect eggs or the peak occurrence of second-to-third-instar larvae. Fungicides should be applied at the early stages of pathogen invasion into plant tissues. It is also recommended to follow the chemical control thresholds for the main diseases and pests in order to achieve better control effectiveness.

(2) *Application technology.* Different pesticides have specific target pests and diseases, and different pests and diseases may respond differently to certain pesticides. Before using crop protection UAVs for pesticide application, it is important to identify the target pests or diseases accurately and use the appropriate pesticides accordingly. Scientifically selecting and matching the pesticide varieties is also an essential step in UAV pesticide application. First, it is important to select efficient, low-toxicity, and low-residue pesticides that specifically target the pests or diseases. Second, the broad-spectrum effect of the pesticide should be considered, allowing for the treatment of multiple pests or diseases with a single pesticide or a proper combination and mixture of several pesticides. The goal is to achieve effective control of multiple pests and diseases with a single application, reducing the number of pesticide applications as much as possible. Additionally, it is important to select pesticide formulations that are suitable for use with UAVs. For pesticide spraying using agricultural UAVs, ultra-low-volume spraying is the main method employed. Among the various pesticide formulations, oil-based formulations (such as emulsifiable concentrates) are preferred for UAV spraying. Suspension concentrates or water dispersible granules can also be used, while commonly used wettable powders are generally more suitable for conventional spraying equipment with constant flow rates. The physical characteristics of the pesticide solution, including viscosity, surface tension, particle size, and evaporation rate, should also be taken into consideration. When spraying pesticides from UAVs, there are significant differences in the environment compared to ground spraying. The physical characteristics of the spray droplets change due to the influence of high-speed airflow, resulting in poor deposition effectiveness. To address this issue, pesticide adjuvants can be used in aerial spraying to reduce droplet evaporation rate, decrease surface tension, and improve droplet size uniformity, thereby promoting droplet deposition. Variable-rate application is expected to be one of the future trends in crop protection technology. Compared to traditional spraying techniques, variable-rate application allows the application of pesticides according to the real-time information on pest density, severity of the infestation, crop density, and other factors. This approach not only solves the problem of excessive pesticide use but also saves costs, improves efficacy, and avoids blind spraying, thus preventing environmental pollution.

(3) *Obstacle-avoidance technology.* The environment surrounding agricultural fields during crop protection UAV operations is generally complex, with structures like buildings and utility poles. These obstacles have a significant impact on safe operations. Therefore, it is important to explore autonomous obstacle-avoidance techniques for UAVs. The challenge lies in the autonomous recognition of obstacles and how to navigate around them, given the diverse shapes and random distribution of obstacles in the vicinity of agricultural fields.

(4) *Drift reduction technology.* There are several factors that affect the drift of spray droplets in crop protection UAV operations, including spraying techniques, canopy types, flight height, weather conditions (wind speed, wind direction, temperature, and humidity), and physical and chemical properties of the pesticide solution. Among these factors, natural wind is the primary factor affecting droplet deposition. Currently, the main approaches for reducing drift include developing drift reduction nozzles, using drift-reducing adjuvants, optimizing spraying techniques, setting up drift buffer zones, and planting hedgerows. The most widely used technique is the use of drift reduction nozzles developed using jet technology. Nozzles are critical components of UAV spraying systems, as they carry and disperse the pesticide solution while reducing droplet drift. Therefore, UAV spraying systems have high requirements for nozzle performance.

(5) *Detection technology.* The evaluation of spray droplets is important for assessing the quality of pesticide application in UAV spraying. There are currently two commonly used methods for evaluation: droplet density determination and 50% effective deposition determination. The 50% effective deposition determination is more recommended as specialists believe that errors may occur when droplets overlap.

(6) *Operational techniques.* The use of crop protection UAVs requires a professional team and skilled operators. Professional technicians must have a comprehensive understanding and flexible application of operational parameters, including the type and quantity of pesticides, adjuvants, flight route planning, flight height, speed, wind direction, droplet size, and more. Accurate settings of these parameters are crucial to ensuring uniform pesticide application during UAV spraying, as well as the overall efficiency and effectiveness of crop protection UAV operations.

4. Application effectiveness of crop protection UAVs

In 2018, the Agricultural and Rural Bureau of Taoyuan County, Hunan Province conducted an experimental demonstration of using crop protection UAVs to control major pests and diseases in rice crops. The trial took place in Xianrenxi Village, Lingjintan Town, and covered a continuous planting area of 66.7 ha for the late-season rice. The "Jifei P20" crop protection UAV was selected for the experiment, with each mission carrying 8 kg of pesticide solution and covering an area of 0.67 ha. Two rounds of ultra-low-volume spraying were conducted. The first aerial spraying for pest control was conducted on August 6, targeting rice planthoppers, Cnaphalocrocis medinalis, and sheath blight. The following pesticide formulations were used: 40% chlorpyrifos ·thiamethoxam water dispersible granules, 50% pymetrozine water dispersible granules, and 30% difenoconazole emulsifiable concentrate. The second aerial spraying for pest control took place on September 3, targeting rice stem borers, rice planthoppers, Cnaphalocrocis medinalis, and sheath blight. The following pesticide formulations were used: 6% avermectin · chlorfenapyr suspension concentrate, 50% pymetrozine water dispersible granules, and 30% propiconazole suspension concentrate. Meanwhile, following conventional spraying methods, local farmers were arranged to manually spray using the Dongfang-hong DFH-16A backpack sprayers with the same pesticides as the aerial spraying. The effectiveness of the pest control was evaluated 15 days after each of the two sprayings. The experimental results showed that the average control efficacy of the "Jifei P20" crop protection UAV for controlling rice stem borers, rice leaf rollers, rice planthoppers, and rice bacterial leaf blight was 87.8%, 92.3%, 82.1%, and 82.9%, respectively. In comparison, the average control efficacy of the common spraying methods applied by local farmers, following their traditional practices, was 86.8%, 92.4%, 87.1%, and 83.9%, respectively. Although the differences between the two methods are not significant, both had achieved the desired control efficacy, and the use of UAVs for crop protection saves labor, costs, and pesticide usage. In 2017, the Agricultural Technology Extension Station of Changxing County, Zhejiang Province, along with other organizations, established a large-scale aerial spraying trial for rice pests and diseases in Houfen Village, Shuikou Township. The Zhejiang "Nongfeike" crop protection UAV was selected for two rounds of pesticide application. During the first round in mid-August, the UAV sprayed a combination of 40% chlorpyrifos · thiamethoxam water dispersible granules, 30% propiconazole suspension concentrate, and pyraclostrobin water dispersible granules. For the second round in mid-September, the UAV sprayed a combination of 6% avermectin · chlorfenapyr suspension concentrate, propiconazole EC, and pyraclostrobin water dispersible granules. Eight days after each spraying, the effectiveness of pest control was assessed. The results showed that the control efficacy for rice planthoppers was 84.07%, while for the third and fourth generations of rice leaf rollers, it was 72.73% and 94.19%, respectively. After evaluating the control efficacy following the prevalence period, it was found to be 89.07% for rice bacterial leaf blight and 83.45% for rice blast disease. In 2018, the Plant Disease and Pest Monitoring and Forecasting Station of Liucheng County, Guangxi, conducted a demonstration of using crop protection UAVs to control rice planthoppers in Dacun Tun, Mashan Town. In mid-June, the "Feiyan" F-10 crop protection UAV was used for one round of spraying. The experimental treatments included five different pesticide formulations: 30% thiamethoxam suspension concentrate, 20% chlorpyrifos EC, 50% buprofezin · chlorpyrifos wettable powder, 25% pymetrozine wettable powder, and 11.5% pyriproxyfen · chlorpyrifos emulsifiable concentration. The control efficacy against rice planthoppers was evaluated 15 days after the spraying and compared with the control area. The results showed that the control efficacies against rice planthoppers were 86.36%, 84.00%, 85.24%, 88.92%, and 85.12% for the five different pesticide formulations, respectively. The differences in control efficacy among these five agents were not significant. Specifically, the control efficacy of the 30% chlorpyrifos suspension concentrate against rice planthoppers was 90.64%, 96.13%, 97.83%, 95.82%, and 86.36% at 1, 3, 7, 10, and 15 days after the spraying, respectively. The efficacy reached its peak at 7 days after spraying and then declined. These medications have achieved similar results to traditional manual application in the past, with good rapidity and durability, which can be widely applied in production.

CHAPTER 10

Case Analysis of Ultra-high Yield of Super Hybrid Rice

I. Key Technologies for Achieving Super-High Yield in Gejiu City, Yunnan Province

1. Basic information

The super hybrid rice demonstration base of Gejiu City is located in the Honghe Hani and Yi Autonomous Prefecture of Yunnan Province, at the southern end of the Yunnan–Guizhou Plateau, with coordinates ranging from N 23° 01' to 23° 36', E 102° 54' to 103° 25'. It covers an area of 1,587 km^2, with an average annual temperature of 16.4°C. The coldest month (January) has an average temperature of 10.1°C, while the hottest month (July) is 20.5°C. The highest altitude is 2,740 m, the lowest altitude is 150 m, and the urban area is at an altitude of 1,688 m. Since the beginning of the twenty-first century, the Gejiu rice base has been conducting high-yielding tackling demonstrations for nearly 20 years. In 2018, the average yield per *mu* reached 1,152.3 kg in the area.

2. High-yield cultivation techniques

(1) *Seedling cultivation.* In dry seedling cultivation methods, such fields should be selected in high terrain for better drainage, with fertile and loose soil and slightly acidic pH. The effect of topdressing in dry seedling cultivation is poor, so attention should be paid to the soil preparation of the seedbed, mainly using decomposed organic fertilizer and farmyard manure, combined with the application of compound fertilizer. Sowing in early March, with a seeding amount of about eight kg/*mu*. Sowing evenly can nurture robust seedlings.

After sowing, insert a bamboo strip every one m on the seedbed as an arch, then cover it with plastic film. Press the edges firmly with soil to seal it for insulation and prevent it from being blown open by rain. Before rooting, pay attention to sealing and insulation to promote rooting and emergence. After the seedlings are rooted and grown, ensure good ventilation to prevent water shortage and diseases caused by high temperatures. After the two-leaf-one-bud stage, uncover the film when the weather is good, and then spray pesticides such as hymexazol and 0.5% urea solution to prevent diseases and promote vigorous growth of the seedlings.

(2) *Transplanting.* It is recommended to transplant the seedlings at around 40 days of age, with an expected leaf age of 5.5 leaves. It is advisable to transplant with wide rows and narrow plant spacing, with a row spacing of 13.3 cm × 30 cm. Each hole should be planted with one grain of rice seedling. Pay attention to planting shallowly (2–3 cm). Shallow planting is an important factor for early tillering and vigorous growth. If planted too deep (>3 cm), a large number of effective tillers at the lower end and several effective secondary tillers that can be produced above will be lost, resulting in a significant reduction in the number of effective panicles per plant, directly leading to a substantial decrease in yield. After transplanting, deep-water irrigation can be carried out for 3–5 days, with the water depth not exceeding the heart leaves. Pay attention to digging productive trenches (about one trench per 0.5 *mu*). For facilitating drainage, the trenches should be wide and deep so that the whole field can be exposed to the sun.

(3) *Fertilizers and application rates.*

a) Field base fertilizer. Apply 500 kg of farmyard manure, 50 kg of basic calcium fertilizer, and 40 kg of compound fertilizer per *mu*, evenly incorporating them into the plowed soil layer when preparing the field. For red soil fields, the amount of farmyard manure can be increased, and a certain amount of lime can be applied to improve the physical and chemical properties of the soil.

b) Tillering fertilizer. Five to seven days after transplanting, apply 20 kg of carbon, 5 kg of urea, and 5 kg of potassium fertilizer per *mu*. Around 15 days after transplanting, assess the seedling growth and apply 3–5 kg of urea and 5 kg of potassium fertilizer per *mu*. Considering the acidity of red soil and the slow growth of seedlings in the early stage, the proportion of ammonium bicarbonate can be increased during the top-dressing process. First, ammonium bicarbonate is alkaline, which can neutralize the acidity of the soil. Second, ammonium bicarbonate has a quick fertilizing effect, which can promote the rapid growth of seedlings.

c) Panicle fertilization. Apply in two separate times, at the third-to-last-leaf stage and the second-to-last-leaf stage. At the third-to-last-leaf stage, apply 5–7 kg of urea, 10 kg of compound fertilizer, and 10 kg of potassium fertilizer per *mu* to promote the healthy development of spikelets. At the second-to-last-leaf stage, apply 3–5 kg of urea and 7.5 kg of potassium fertilizer per *mu* to prevent spikelet degradation and lay a good foundation for the formation of large spikes.

d) Grain fertilization. During the heading stage, spray 100 g of grain saturates diluted in 60 kg of water per *mu* onto the leaves to reduce the percentage of empty husks, increase the fruit-setting rate, and improve grain weight.

(4) *Field-drying for seedling control.* Begin draining water and drying the field when the number of seedlings reaches 80% (150,000) of the planned panicle number of 180,000. If the number of seedlings is low earlier in the season, start field-drying when it is about 90% of the planned panicle number. The method involves multiple light drying sessions after watering (ensuring that each drying session gets to the point where a person can stand in the field without their feet sinking, and for red loam fields, the drying time can be appropriately increased each session) to induce leaf yellowing (the color of the top three leaves should be darker than the top four leaves). This process can be continued until the early stage of panicle differentiation. If the leaves have not faded in color by the specified leaf age, continue with the drying process and refrain from re-watering and fertilization.

(5) *Water management.* Super rice combinations have well-developed root systems and strong growth potential. To encourage early growth, control ineffective tillering toward the end of the tillering phase, and ensure root vitality later, the focus of water management is on "enhancing aeration, nurturing roots, and maintaining vitality." The specific methods include maintaining shallow water after transplanting and regreening of seedlings, promoting tillering with moist irrigation during the tillering stage, and starting to sun-dry the field when the number of seedlings reaches 80% of the targeted panicle number. Multiple-time mild sun-drying sessions are employed to control the ineffective tillering, encouraging root growth and deep penetration. Excessive sun-drying that enlarges the gaps in the field can cause irreparable root damage. Mild sun-drying is preferable until the leaf color lightens (the color of the third top leaf is darker than the fourth leaf). If the target is not met, mild sun-drying can continue until the early differentiation stage of the young panicles on the main stem. After panicle differentiation, maintain shallow water until the heading and flowering stage, use intermittent irrigation during the grain-filling and maturing phase, alternating between dry and wet conditions, with a focus on moist conditions during the flowering period and dry conditions later to ensure root vitality, prevent premature aging, and improve the setting rate and fullness. Ensure alternating dry and wet conditions, maintain clear water and firm panicles, nurture with air for roots, protect leaves with roots, and increase weight with leaves. All these actions aim to achieve the required effective panicle number for high-yielding and ensure full maturity and high productivity.

(6) *Pest and disease control.* During the seedling stage, pest control treatments are applied one to two times, primarily targeting rice thrips and rice gall midge. One day before transplanting, the nursery is sprayed with pesticides once to reduce the incidence of pests and diseases in the early stage of the field growth. To mitigate pesticide residues and promote low-carbon cultivation, the application of pesticides in this period is mainly preventive, with two to three treatments targeting three pests and three diseases: the rice stem borer, rice leaf roller, rice planthopper for pests; and

rice blast, sheath blight, and bacterial leaf blight for diseases. The timing for pesticide application is determined based on field surveys.

II. Key Technology for Achieving Super-High Yield in Longhui County, Hunan Province

1. Basic information

1) *Base information.* Longhui County is located in the eastern part of the Xuefeng Mountain range. Niuxing Village in Yanggu'ao Township is situated in the northeastern part of the county, in a low to mid-elevation mountain area, with an altitude of 370 m. The climate is mild, with ample sunshine and heat, and abundant rainfall. The average annual temperature is 16.6°C, with the initial day having an average temperature ≥10°C for five consecutive days falling between April 2 and 6. The annual rainfall is around 1,330 mm, and the frost-free period is approximately 270 days. The soil has an organic matter content of 35.2 g/kg. The total amounts of nutrients N, P_2O_5, and K_2O are 1.58 g/kg, 0.50 g/kg, and 2.94 g/kg, respectively. The contents of alkali-hydrolyzable nitrogen, available phosphorus, and available potassium are 124 mg/kg, 5.1 mg/kg, and 165 mg/kg, respectively. The pH value is 5.5. The research base has sandy loam soil, with a deep soil layer, strong water and fertility retention capabilities, complete drainage and irrigation facilities, and favorable ecological conditions. This research site has been undertaking high-yield research tasks since the late 1990s and is a demonstration base with a strong technical foundation and popular support.
2) *Research combination and yield performance.* The ultra-high-yielding hybrid combination Y Liangyou 900 was developed under the guidance of the new super hybrid rice breeding research techniques proposed by Yuan Longping. It features the typical ideal plant architecture of super rice, with a high canopy layer and a short panicle layer. The plant's canopy morphology is characterized by long, straight, narrow, concave, and thick leaves. The panicle layer is relatively short. After heading and grain-filling, the overall center of gravity of the plant is lowered, making it less prone to lodging. The variety has a plant height of 129.5 cm, with a total of 15 leaves on the main stem and 6 elongated internodes. It produces about 2.28 million effective panicles per ha, with a panicle length of 31.7 cm. Each panicle has approximately 340.2 grains, with a grain-setting rate of 93.6%, and a thousand-grain weight of 27.5 g.

2. High-yield cultivation techniques

Based on the characteristics of Y Liangyou 900, the spikelet structure design for the demonstration field aims for 2.55 million effective panicles per ha, with an average of 300 grains per panicle, a seed setting rate of 90%, and a thousand-grain weight of 27.5 g.

1) *Seedling cultivation and transplantation.* Seed soaking begins in mid-April, with seed sunning, selection, and disinfection treatments conducted before soaking. Sprouted grains are then cultivated in water for seedling growth. After plowing, a base fertilizer of 40 kg of 45% (15-15-15) compound fertilizer is applied per *mu*, with a seeding rate of 7.5 kg/*mu* in the nursery field and approximately 1 kg/*mu* in the main field, both sown sparsely to foster strong seedlings. Before transplanting, the main field is finely plowed with a medium-sized plowing machine to a depth of 30 cm, then leveled to ensure a 3 cm layer of water without exposing mud. Based on the yield design and characteristics of Y Liangyou 900, the transplantation adopts a wide and narrow row spacing, with the wide rows being 40 cm apart and the narrow rows 23.3 cm apart, and a plant spacing of 20 cm, equating to about 10,500 plants per *mu*. The wide rows are oriented east-west, and a specialized row marker is used for line marking and transplantation, which benefits ventilation and light penetration in the middle and late stages.
2) *Rational fertilization.* For a target yield of kg/*mu*, each 100 kg of rice grains requires 1.8 kg of nitrogen; thus, the total nitrogen requirement is $1.8 \times 10 = 18$ kg/*mu*. Assuming the soil supplies 8 kg of pure nitrogen per *mu*, the additional required pure nitrogen is $18 - 8 = 10$ kg/*mu*. A fertilizer utilization rate of 40% means that 25 kg of nitrogen per *mu* needs to be applied additionally. The proportion of fertilization before and after is set at base tillering fertilizer to panicle fertilizer in a ratio of 6:4, meaning 15 kg/*mu* of pure nitrogen for base tillering fertilizer and 10 kg/*mu* of pure nitrogen for panicle fertilizer. Besides applying nitrogen fertilizer, high-yielding cultivation should also focus on the rational application of phosphorus and potassium fertilizers to achieve balanced fertilization. The general ratio of nitrogen, phosphorus, and potassium should be 1:0.6:1.1, with the fertilizers applied in three categories: base fertilizer, tillering fertilizer, and panicle fertilizer.
3) *Scientific water management.* After transplanting, irrigate with shallow water and pause irrigation from the revival and liveliness phase to the critical tillering phase. When about 80% of the expected number of seedlings (approximately 140,000 per *mu* out of 170,000 per *mu*) is reached, start draining water and drying the fields using multiple-time sun-drying methods until the leaf color fades. During the grain-filling to maturing stage, employ intermittent irrigation with dry-and-wet alternation, focusing on wet conditions during the flowering phase and dry conditions in the later stages, which is to ensure root vitality, prevent premature aging, increase the seed setting rate, and enhance grain-filling.
4) *Integrated disease and pest management.* Disease and pest management should primarily focus on prevention, combining prevention and control under a unified management principle. Key targets include rice thrips during the seedling phase in the nursery, rice leaf roller, stem borers, rice planthoppers, rice blast, and sheath blight during the main field phase.

Particularly under high-fertility and high-density conditions, special attention should be paid to sheath blight and rice planthoppers in the middle and late stages. Based on pest forecasting, integrated management should be conducted using chlorantraniliprole (Coragen) for controlling rice leaf roller and stem borer, tolfenpyrad for controlling rice planthoppers, validamycin for rice blast prevention, and isoprothiolane for treating sheath blight.

III. Key Technologies for Achieving Super-High Yield in Guangshan County, Henan Province

1. Basic information

Guangshan County is located in the southeastern part of Henan Province, at the junction of the provinces of Henan, Hubei, and Anhui. It is bordered by the Huai River to the north and the Dabie Mountains to the south. The county stretches 60 km from east to west and 55 km from north to south, covering a total area of 1,835 km^2, with a population of 860,000 people. The total water resources of the county amount to 2 billion m^3, with a forest coverage rate of 42.6%. The county currently has 1.37 million *mu* of cultivated land. The National Hybrid Rice Engineering Technology Research Center selected Guangshan County in Henan for the implementation of the "Super Hybrid Rice 'Hundred, Thousand, and Ten Thousand' High-Yielding Tackling Demonstration" project from 2014 to 2016. The average yield per *mu* in the ten thousand *mu* demonstration base exceeded 800 kg, in the thousand *mu* demonstration base exceeded 900 kg, and in the hundred *mu* tackling field, the yield surpassed 1,000 kg per *mu*. This research led to the development of high-yielding technical specifications for super hybrid rice over large areas.

2. High-yield cultivation techniques

1) *Seedling cultivation.* For hand-transplanted seedlings, use either the small arch-shed moist-and-sparse sowing method for robust seedling cultivation or the arch-shed tray-broadcasting method for seedling cultivation.

    a) Sowing time and volume. The sowing period depends on the timing of the previous crop, with sowing in mid-April and the seedling age controlled at around 25 days. The sowing amount for each *mu* of the seedbed is 10 kg, and each tray for broadcasting seedlings should have 40–50 g of dehulled grains. Seed by dividing the tray and weighing, with about 40 trays needed per *mu* of the main field. Sow 70% first and reserve 30% for replanting to ensure even seeding. The weight of seeds used per *mu* of the main field is 1 kg.

b) Seed soaking and germination stimulation. Before soaking the seeds, sun-dry the seeds for two sunny days, then wash and float them to remove the less viable grains. Soak the seeds in thiram or strong bleach solution for 24 hours, strictly following the medication operation procedures. Then, rinse the seeds thoroughly and combine them with the "three up, three down" method for soaking and germination stimulation. When 80% of the seeds have cracked open, spread them out at room temperature for six hours, and then they are ready for sowing. If sowing cannot be done in time, spread out the germinated seeds. When the surface of the seeds is dry, spray a small amount of warm water.

c) Selection and preparation of the seedbed. Select seedbeds that are near water sources, easy irrigation and drainage, sheltered from the wind but facing the sun, and easy to manage, and obtain deeply plowed soil, such as old rice fields, vegetable gardens, or winter fallow paddy fields. Before leveling the seedbed, apply 5–6 kg of matured farmyard manure per *mu*, 50 kg 40% specialized fertilizer (or specialized seedbed fertilizer), 8 kg potassium chloride, 1 kg zinc fertilizer, and 6 kg/*mu* silicon fertilizer. Thoroughly till the soil. For water-leveled seedbeds, wait until the mud settles before making the seedbed, and for dry-leveled seedbeds, wait 4–5 days before forming the seedbed. Typically, the seedbed should be around 1.5 m wide (or based on the width of the seedling tray or film), no longer than 15 m, with a spacing of 0.7 m between beds and a distance of 1 m at the head of the bed.

d) Seed sowing with film covering. During sowing, it is necessary to divide the beds and weigh them accurately, distribute the seeds evenly, and lightly compress them to ensure close contact between the seeds and the soil or embed the seeds on three sides in the mud, then cover them with prepared fine soil (sieved fine soil mixed with 20%–40% well-rotted farm manure) to a thickness of 0.8–1 cm. For tray nursery preparation, divide and weigh each tray, fill the tray soil up to 2/3, evenly sow the seeds, then cover with soil, and then use a wooden board to scrape off the excess nutrient soil. For each *mu* of the nursery bed, use 3% Kasugamycin (40 g mixed with 40 kg of water) for spraying and disinfecting the seedbed.

e) Seedbed management. First, temperature. After sowing until before rooting, it is important to keep the seedbed sealed and warm to promote rooting. The favorable temperature is 30°C–32°C. If the temperature exceeds 38°C, appropriate ventilation should be used to cool down to prevent the seeds and buds from burning. After 60% of the seedlings turn green, a slight ventilation should be used to harden the seedlings, reducing the inside temperature to about 25°C. The temperature cannot exceed 35°C, if not, the seedlings burned. After the one-leaf-one-bud stage, the film can be removed in good weather conditions. After removing, spray pesticides such as hymexazol and 0.5% urea solution to prevent diseases and promote the robust growth of the seedlings. Depending on the occurrence of pests and diseases, spray one to two times with pesticides such as Regent and Fuji No. 1 in the seedling fields to prevent and control pests and diseases like rice thrips, striped stem borer, rice water weevil, sheath blight, and rice blast, ensuring the robust growth of seedlings without pest or disease damage. Second, topdressing. Apply weaning fertilizer when the seedlings are in the two-leaf-one-bud stage. Every 10 m^2 sprays the mixture of 25 g of urea and 5 kg of water. It is essential to spray clean water to wash the seedlings after fertilizing. Apply the last time fertilizer five to seven days before transplanting, with 3–5 kg of urea per *mu* to promote regreening and revitalization of the plants. Also, attention should be paid to the occurrence and control of pests and diseases such as rice stem borers. Third, water management. From seeding to the stage before the one-leaf-one-bud stage, try to keep the seedlings growing in conditions close to dry fields. Pay attention to the water irrigation criteria for the seedbed; irrigate thoroughly once in the morning on a sunny day if any of the following conditions are observed: absence of dew on the seedlings in the morning and evening, dry soil in the seedbed, or leaves curling at noon. From the one-leaf-one-bud stage to the three-leaf-one-bud stage, fill the ditches with water to keep the seedbed moist. After the three-leaf stage, each irrigation should add a 0.5–1 cm layer of water and let it dry naturally for two to three days. Repeat this cycle several times. Maintain a water layer of about 1 cm under the conditions of fertilization, weeding, pest and disease control, and intense sun exposure that require water.

2) *Transplanting.*

a) Transplanting time. For sparse-planting cultivation for strong seedlings, transplant them in a timely manner at the four-to-five-leaf stage with soil; for broadcasting-trays cultivation, transplant them at the two-and-a-half-to-three-leaf stage with soil.
b) Field preparation. Before transplanting, perform precise field preparation on the main field. Use a medium-sized plowing machine to deep plow the rice field to about 30 cm, then level it, making sure mud can be seen in 3 cm layers of water. Based on the yield targets, local climate conditions, and the characteristics of the variety, planting strategies of wide-and-narrow rows or wide rows with narrow plantations should be adopted. The rows face east-west, and the spacing should be 30 cm × 16.7 cm, with 13,300 holes can be inserted per *mu*. Each hole should have two strong seedlings with tillers. During transplantation, it is essential to insert shallowly (2–3 cm) because shallow planting is a prerequisite for more and earlier low-position tillering, ensuring enough seedlings and large panicles. To secure a sufficient number of seedlings, transplant them transplant immediately after pulling with soil attached, inserting them securely in a straight position. Do not transplant weak or diseased seedlings, avoid excessive or missed insertion.

3) *Fertilization.*

a) Base fertilizer. Combined with field cultivation, apply 50 kg of rapeseed cake fertilizer per *mu*, 40 kg of 40% rice-specific fertilizer (sulfur-based), 40 kg of calcium superphosphate, 12 kg of potassium chloride, 1 kg of zinc sulfate, 6 kg of silicon fertilizer, and 1 kg of borax per *mu*.
b) Tillering fertilizer. Five to seven days after transplanting and once the seedlings have established, apply 8–10 kg of urea per *mu*. Thirteen days after transplanting, depending on the condition of the seedlings, apply an additional 3 kg of urea and 6 kg of potassium chloride per *mu*, ensuring balanced fertilization (avoid nutrient deficiencies and further fertilization).
c) Panicle fertilizer. Apply it twice. After the fields have been sunned and the collective leaf color turns yellow, the first-time application should be carried during the young panicle differentiation 2–3 stages: 10 kg of 45% (15-15-15) compound fertilizer, 7 kg of urea,

and 13 kg of potassium chloride per *mu*. If the discoloration happened in the first stage, apply in the second stage; if discoloration is in the second stage, apply in the fourth stage; and if discoloration is in the third stage, apply in the third stage. For fields showing earlier discoloration due to nutrient deficiency, compound fertilizer can be counted into urea for application. The second application should be carried out during the fourth stage of panicle differentiation, applying 3 kg of urea and 6 kg of potassium chloride per *mu*.

d) Grain fertilizer. During the heading stage, apply one packet (100 g) of grain satiety mixed with 60 kg of water per *mu* for foliar application; during the grain-filling stage, apply microelement fertilizers and potassium dihydrogen phosphate one to two times as a foliar spray.

4) *Water management.*

a) Thin water for transplanting, shallow water for regreening. Keep a thin layer of water during the transplanting to ensure the quality of transplanting, for deep water will cause floating and missing seedlings. Irrigate with a water layer of about three cm within five to six days after transplanting to create a stable environmental condition in terms of temperature and humidity, which will promote the generation of new roots and the rapid rejuvenation and revival of the plants.
b) Shallow water and moist conditions for tillering. Implement intermittent irrigation from the period of rejuvenation to the critical period of effective tillering should maintain dry and wet alternation with a focus on wet conditions. Combined with intervillage weeding and topdressing, irrigate with a water layer of one to two cm until it naturally dries out for three to four days, and then repeat the process. No muddy water during sunny days and no water layer on rainy days. This will promote root growth, cause earlier tillering, and lower the tillering node position.
c) Gentle sun-drying for robust seedlings. When the total number of seedlings per *mu* reaches 80% of the planned number of panicles per *mu* (around 150,000 per *mu*), start draining the field for sunning. Adopt the method of multiple gentle sunning. Generally, Sun the field until it is slightly cracked and white roots on the field surface, no sinking when people step on it, and standing leaves with their color faded. If the color of the leaves has not faded when it comes to the stage of applying spike fertilizer, continue sun-drying the field and do not apply panicle fertilizer.
d) Water nourishment during panicle development. Resume irrigation when the main stem of the rice plant is in the 2–3 stages of young panicle differentiation. Adopt shallow, frequent watering and then naturally drying out. Then re-irrigate timely after the mud shows for one to two days. Around the period before and after the meiotic division of panicle differentiation (panicle differentiation phases 5–7), maintain a continuous water layer of three to four cm without drying out.
e) Sufficient water during heading. Maintain a shallow water layer during the heading and flowering stage to create high humidity, which is beneficial for normal heading and pollination.
f) Alternative dry-and-wet irrigation for robust grains. From the late flowering stage until maturity, alternate between dry and wet conditions focusing on wetness. This practice enhances

root vitality, and delays root aging, so roots can get more oxygen. Robust roots will result in more leaves and then more grains.

g) Dehydration when fully ripened. Drain and then sun-dry fields five to seven days before harvesting when the crop is fully ripened. Do not cut off water too early, otherwise it will affect seed-filling and yielding.

h) Dig the field surrounding ditches at the second application of panicle fertilizer and dig cross-shaped ditches within the field. The standard size for these ditches, both surrounding and within fields, is about 30 cm deep and 20 cm wide. For every area larger than 333 m^3, dig a compartment ditch that is approximately 20 cm wide and 20 cm deep. These ditches are beneficial for field drainage and irrigation or for drying the field.

5) *Implement comprehensive pest, disease, and weed management.* Pests and diseases favor high fertility and dense planting. The plant protection strategy should focus on prevention and integrated control. Attention should be given to the comprehensive prevention and control of pests and diseases such as rice thrips, rice stem borers, rice leaf rollers, rice planthoppers, blast disease, sheath rot, sheath blight, and false smut. Chemical control recommendations include the following pesticides:

a) For rice thrips, rice borers, and rice leaf rollers, use chlorantraniliprole (Coragen), Daoteng, profenofos, chlorpyrifos, avermectin, and emamectin benzoate.

b) For planthoppers, use buprofezin, pymetrozine, and chlorpyrifos.

c) For rice blast, first disinfect seeds through soaking. Second, add an appropriate amount of fungicide for spraying during key stages. Third, if sporadic outbreaks occur, control in time with Fuji No. 1, tricyclazole, or kasugamycin. Fourth, in severe conditions, apply fungicides continuously every five to seven days until eradicated. For seedling blast, disinfect seeds, focus on controlling seedling blast during the three-leaf stage, and conduct general prevention three to five days before transplanting. For leaf blast, implement comprehensive control 15 to 20 days after transplanting and use tricyclazole for leaf blast prevention. If the leaf blast is detected, apply the fungicide timely. If severe, repeat the treatment two to three times at an interval of one week. For neck blast, spray tricyclazole one to three days before heading for prevention; spray ethaboxam at heading; if symptoms are detected, apply 6% kasugamycin 80 ml per *mu* until eradicated.

d) For sheath blight, use products like armure, tebuconazole, and jinggangmycin.

e) For rice dwarf disease, it should be notified that the occurrence of southern rice black-streaked dwarf virus has intensified in recent years, leading to substantial crop losses. This disease, transmitted by planthoppers, should be taken seriously. Control measures should focus on the nursery and early transplanting stages to prevent disorder and dwarfism, aiming for "controlling disorder to prevent dwarfism, controlling pests to prevent disease."

f) For rice false smut, use jinggangmycin, triadimefon, carbendazim WP, etc., should be used during the late booting stage (five to seven days before heading). If a second-time control is needed, apply during the heading stage (around 50% heading); prevention at the full heading stage is less effective.

Bibliography

Ai, Zhiyong, Guo Xiayu, and Liu Wenxiang, et al. "Changes of Safe Production Dates of Double-Cropping Rice in the Middle Reaches of the Yangtze River." *Acta Agronomica Sinica*, no. 07 (2014): 1320–1329.

Bai, Hongsong, Feng Zhichun, and Ren Yuanli. "Study on the 'Three Super Cultivation' Rice Varieties and Transplanting Specifications." *Reclamation and Rice Cultivation*, no.5 (2005): 15–17.

Bi, Xu. "Analysis of Changing Characteristics of Temperature and Rainfall in Hubei Province and Their Correlation with Geographical Factors." PhD diss., Wuhan: Central China Normal University, 2013.

Chi, Zaixiang, Yang Guilan, and Yang Li, et al. "Study on the Effects of Light and Temperature Factors at Different Altitudes on the Yield of Super Rice Liangyou 106." *Journal of Guizhou Meteorology* 31, no. 6 (2007): 9–10.

Chu, Chenghu, and Zhang Huizhen. "Research on High-Yielding Key Technologies of Super Rice in Low-Altitude Areas of Dabie Mountains." *Modern Agriculture Science and Technology*, no. 7 (2015): 41, 45.

Dong, Danhong. "Spatio-Temporal Analysis of Changes in Temperature with Altitude in China in a Global Context." PhD diss., Chengdu: Chengdu University of Information Technology, 2015.

Gu, Ming. "Effects of Altitude on Rice Growth and Development." *Tillage and Cultivation*, no. 1/2, 1997: 61–63.

He, Wenhong, Chen Huizhe, and Zhu Defeng, et al. "Effects of Different Planting Rates on the Quality and Yield of Rice Machine-Transplanted Seedlings." *China Rice*, 2008, 14 (3): 60–62.

Huang, Qiansheng. "Effects of Planting Density of Rice on the Dynamics and Yield of Agricultural Machinery." *Agricultural Machinery Service* 30, no. 6 (2013): 558–559.

Huang, Zhinong, and Zhang Yuzhu. "The Theory and Practice of Ecological Regulation of Rice Pests." *Crop Research* 20, no. 4 (2006): 297–307.

Huang, Zhinong, Liu Yong, and Zhang Ling, et al. "Preliminary Exploration on the Occurrence and Control of Lissorhoptrus Oryzophilus kuschel in Hunan Province." *Hunan Agricultural Sciences*, no. 2 (2006): 62–63.

Huang, Zhinong, Ma Guohui, and Xu Zhide, et al. "Effects of High-Yielding and Nitrogen-Saving Cultivation of Hybrid Rice on the Occurrence of Damage from Chilo Suppressalis." *Modern Agricultural Science and Technology*, no. 8 (2010): 159–161.

Huang, Zhinong, Ma Guohui, and Zeng Xiaoling, et al. "Effect of Fertilizer Management on the Occurrence of Chilo Suppressalis in Hybrid Rice." *Hybrid Rice* 25, no. 5 (2010): 76–79.

Huang, Zhinong, Xu Zhide, and Wen Jihui, et al. "Study on the Application and Control Effect of Sex Attractant of Chilo Suppressalis in Rice Production." *Hunan Agricultural Sciences*, no. 10 (2009): 64–69.

Huang, Zhinong, Zhang Yuzhu, and Liu Yong, et al. "Causes and Control Strategies of Three Major Rice Pests in Hunan." *Crop Research* 20, no. 4 (2006): 315–317, 323.

Huang, Zhinong, Zhang Yuzhu, and Zhu Guoqi, et al. "Evaluation of the Control Effect of Trichogramma Japonicum on Cnaphalocrocis Medinalis and Chilo Suppressalis." *Acta Agriculture Jiangxi* 24, no. 5 (2012): 37–40.

Huang, Zhinong. "Green Prevention and Control Practical Technology of High-Quality Rice Pests and Diseases." *Hunan Agricultural Sciences*, no. 4 (2012): 5–9.

Huang, Zhinong. "Green Prevention and Control Strategy of High-Quality Rice Pests and Diseases." *Hunan Agricultural Sciences*, no. 14 (2011): 11–12.

Huang, Zhinong. *Integrated Management of Diseases and Pests in Hybrid Rice*. Changsha: Hunan Science and Technology Press, 2011.

IPCC. *Climate Change 2014 Synthesis Report*. Geneva, Switzerland, 2014.

Jin, Chuanxu, Zhong Qinfu, and Huang Daying, et al. "Effects of Planting Density and Number of Seedlings Planted in Holes on Rice Yield and Its Components." *Guizhou Agricultural Sciencesi* 40, no. 4 (2012): 85–87, 90.

Kong, Xuemei, Wu Fei, and Wang Jun, et al. "Effects of Two-Night Hybrid Rice Planting Density and Number of Seedlings Planted in Holes on Yield." *Modern Agricultural Technology*, no. 19 (2014): 12, 14.

Li, Cunxin, and Lin Dehui. "Dry-matter Production and Partitioning in the Aerial Part of Rice Grown at Different Altitude Localities." *Acta Botanica Yunnanica* 9, no. 1, 1987: 89–95.

Li, Ganghua, Zhang Guofa, and Chen Gonglei, et al. "Group Characteristics and Nitrogen Effects of Ultra-High-Yielding Conventional Japonica Rice Ningjing No. 1 and Ningjiong No. 3 Response." *Acta Agronomica Sinica* 35, no. 6 (2009): 1106–1114.

Li, Huizhu, Liu Chaodong, and Fu Rongfu, et al. "Different Planting Numbers and Densities Affected the Dynamics and Yield of the China Airlines Group 31 Effects of Composition." *Guangdong Agricultural Sciences* 44, no. 5 (2017): 1–6.

Li, Jianwu, Deng Qiyun, and Wu Jun, et al. "Analysis on High-yielding Cultural Techniques for Super Hybrid Rice YLY2 Planted in Sanya." *Chinese Journal of Tropical Agriculture*, 2012 (09): 6–11.

Li, Jianwu, Deng Qiyun, and Wu Jun, et al. "Characteristics and High-yielding Cultural Techniques of New Super Hybrid Rice Combination Y Liangyou 2." *Hybrid Rice*, 2013 (1): 49–51.

Li, Jianwu, Long Jirui, and Guo Xiayu, et al. "High-yielding Cultural Techniques for Mid-season Super Hybrid Rice." *Tillage and Cultivation* 40, no. 1 (2020): 61–62.

Ling, Qihong. "Discussion on the Problem of Light and Simple Cultivation of Rice." China *Rice*, no. 5 (1997): 3–9.

Ling, Qihong. *Crop Population Quality*. Shanghai: Shanghai Science and Technology Press, 2000.

Ling, Qihong. *Theory and New Technologies for High-Yield and High-Efficiency Rice Production*. Beijing: China Agricultural Science and Technology Press, 1996.

Long, Jirui, Ma Guohui, and Song Chunfang, et al. "Effects of Nitrogen-Saving Cultivation with Different Fertilizers on Growth, Yield, and Nitrogen Using Efficiency of Super Hybrid Mid-Season Rice." *Research of Agricultural Modernization* 29, no. 1 (2008): 112–115, 127.

Long, Jirui, Ma Guohui, and Song Chunfang, et al. "Study on Nitrogen Application Rate and Nitrogen, Phosphorus, and Potassium Ratio Mode of Super Hybrid Mid-Season Rice Nitrogen-Saving Cultivation." *Research of Agricultural Modernization* 29, no. 4 (2008): 494–497.

Long, Jirui, Ma Guohui, and Xu Wenyan, et al. "Effects of Plant Growth Retardant Lifengling on Lodging Resistance and Yield of Mid-Season Hybrid Rice." *Hybrid Rice*, 2011, 26 (1): 56–60.

Long, Jirui, Ma Guohui, and Zhou Jing, et al. "Effects of Slow-Release Urea on Growth and Nitrogen Using Efficiency of Super Hybrid Rice Y Liangyou No. 1." *Hybrid Rice* 22, no. 6 (2007): 48–51.

Long, Jirui, Ma Guohui, and Zhou Jing, et al. High-Yielding and High-Efficiency Planting Mode and Techniques of Mini Watermelon and Super Hybrid Rice." *Phosphate & Compound Fertilizer*, no. 03 (2008): 68–69.

Long, Jirui, Song Chunfang, and Ma Guohui, et al. "Effects of Mechanical Precision Hill-Direct-Seeding and Positioning Fertilization on Rice Growth and Nutrient Migration." *Hybrid Rice*, 2014, 29 (3): 60–64.

Lü, Tengfei, Zhou Wei, and, Li Yinghong, et al. "Effects of Seedling Age, Water and Fertilizer Management Model, and Number of Seedlings on Stem Material Production and Yield at Different Levels of Indica Hybrids." *Hybrid Rice* 32, no. 1 (2017): 52–61.

Ma, Guohui, Liu Maoqiu, and Luo Fulin. "Comparative Study on the Application Effect of Super Hybrid Rice Special Fertilizer in Hybrid Rice." *Hybrid Rice*, no. 04 (2013): 37–39.

Ma, Guohui, Tang Haitao, and Wan Yizhen, et al. "Comparison and Evaluation of Yield and Nitrogen Absorption and Utilization Efficiency of Late Hybrid Rice under Different Nitrogen Levels." *Hybrid Rice*, no. 03 (2010): 35–38.

Ma, Guohui, Long Jirui, and Dai Qingming, et al. "Study on the Optimum Slow-Release Nitrogen Fertilizer Dosage and Density Configuration of Super Hybrid Rice Y Liangyou No. 1." *Hybrid Rice*, no. 06 (2008): 73–77.

Ma, Guohui, Long Jirui, and Tang Haitao, et al. "Rice Nitrogen-Saving High-Yielding and High-Efficiency Cultivation Techniques Strategy and Practice." *Hybrid Rice*, no. S1 (2010): 338–345.

Ma, Guohui, Xiong Xurang, and Pei Youliang. "On Main Limited Factors and Strategies for High-Yielding Cultivation of Super Rice in Hunan Province." *Hunan Agricultural Sciences*, no. 03 (2005): 23–25.

Ma, Guohui, Zhou Jing, and Long Jirui, et al. "Effect of Slow-released Nitrogen Fertilizer on Growth and Yield of Super Hybrid Early Rice." *Journal of Hunan Agricultural University (Natural Sciences)*, no. 01 (2008): 95–99.

Ma, Guohui. "A Preliminary Discussion of Theory and Practice on High Yield of Super Hybrid Rice." *Journal of Agricultural Science and Technology*, no. 4 (2005): 3–8.

Ma, Guohui. "High-yielding and High-Quality Fertilization Techniques for Hybrid Rice with Helium." *Hunan Science and Technology News*, January 18, 2005.

Ma, Guohui. "Nitrogen-Saving Cultivation Techniques of Super Hybrid Rice." *Hunan Agriculture*, no. 01 (2008): 11.

Ma, Guohui. "Slow-Released Fertilizers Will Promote Nitrogen Reduction and Yield Increase in Rice." *China Agri-Production News*, no. 07 (2008): 30–31.

Ma, Guohui. "Study on Nitrogen-Saving Cultivation of Hybrid Rice by Synchronous Direct Seeding and Deep Fertilization." *Agriculture Machinery Technology Extension*, no. 12 (2010): 14–15.

Peng, Sumei, and Ling Xiaohui. "Analysis of the Promotion, Application, and Effectiveness of Green and Efficient Rice Promotion Technology in Yichun City." *China Agricultural Technology Extension* 35, no. 9 (2019): 34–36.

Piao, S. L., P. Ciais, and Huang Y., et al. "The Impacts of Climate Change on Water Resources and Agriculture in China." *Nature* 467, no. 7311 (2010): 43–51.

Qu, Shiyong, and Guo Lina. "Rice's Water Requirement Rules and Moisture Management Techniques at Various Growth Stages." *Jilin Agriculture* 264, no. 2 (2012): 100.

Shen, Hongchang, Ma Guohui, and Song Chunfang. "Research Progress of Rice Stem Morphology and Lodging." *Hunan Agricultural Sciences*, no. 08 (2009): 41–44.

Shi, Chunting, Huang Yong, and Ye Jianchun, et al. "Effects of the Number of Seedlings Planted in Different Holes on the Growth and Yielding of rice Y Liangyou 900." *Agricultural Science and Technology Bulletin*, no. 8 (2017): 70–73.

Shi, Lijuan. "Study on Dynamic Plant Types of Super Hybrid Rice Y Liangyou No. 1." PhD diss., Changsha: Hunan Agricultural University, 2007.

Shi, Nianzhen, Zhou Shangquan, and Xu Jianguo, et al. "The Role of Rice Quality Safety Control Technology in Pest and Disease Control." *Crop Research* 20, no. 4 (2006): 337–341.

Song, Chunfang, Shu Youlin, and Peng Jiming, et al. "High-yielding Cultural Techniques of Super Hybrid Rice in Large–scale Demonstration with a Yield over 13.5 t/hm^2 at Xupu, Hunan." *Hybrid Rice* 27, no. 06 (2012): 50–51.

Song, Chunfang, Wen Jihui, and Yang Lu, et al. "Comparative Experiment of Early Rice in Precise Rice Hill-Drop Drilling and Manual Broadcast-Sowing." *Hunan Agricultural Sciences*, no. 02 (2016): 16–18.

Song, Chunfang. "Performance and High-yielding Cultural Techniques of Super Hybrid Rice in Demonstrative Production at Longhui, Hunan." *Hybrid Rice* 26, no. 04 (2011): 46–47.

Tan, Yaling, Hong Ruke, and Chen Jinfeng, et al. Study on the Effect of Altitude on the Growth of Different Rice Varieties." *Seed* 28, no. 7 (2009): 27–30.

Tan, Zhengzheng, Wei Zhongwei, and Ma Guohui. "Analysis of Yield and its Component Factors of Medium Indica Rice in Hunan." *Crop Research*, no. 05 (2015): 463–467.

Tian, Xiaohai, Luo Haiwei, and Zhou Hongduo, et al. "History, Progress, and Prospects of Rice Heat Damage Research in China." *Chinese Agricultural Science Bulletin*, 2009 (22): 166–168.

Tian, Xiaohai, Wang Xiaoling, and Xu Fenying, et al. "Effects of Plant Growth Regulator Lifengling on Lodging Resistance and Canopy Structure of Super Hybrid Rice." *Hybrid Rice*, no. 3 (2010): 64–67, 73.

Tian, Xiaohai, Wu Chenyang, and Yuan Li, et al. "Grain-Setting Rates and Their Correlations with Meteorological Factors under Normal Climatic Conditions in Super Hybrid Rice in the Jianghan Plain, China." *Chinese Journal of Rice Science*, no. 05 (2010): 539–543.

Wei, Zhongwei, and Ma Guohui. "Biological Characteristics and Lodging Resistance of Super-High-Yielding Hybrid Rice Chaoyou 1000." *Hybrid Rice* 30, no. 1 (2015): 58–63.

Wei, Zhongwei, and Ma Guohui. "Studies on Characteristics of Root System of Super-high-yielding Hybrid Rice Combination Chaoyou 1000." *Hybrid Rice*, 2016, 31 (5): 51–55.

Wu, Zhaohui, Li Youzhen, and Wang Xiaoning, et al. "The Performance and High–yielding Cultivation Techniques of Y Liangyou 2 in the 100-*mu* Demonstrative Production at Chengmai, Hainan." *Hybrid Rice* 26, no. 6 (2011): 51–53.

Wu, Zhaohui, Yuan Longping, and Qing Xianguo. "History and Progress of Research on Super High-Yielding Rice Breeding." *Journal of Hunan Agricultural University (Natural Sciences)* 34, no. 1 (2008): 1–5.

Xu, Fei. "Effects of Different Planting Densities on the Growth and Yield of Different Rice Varieties." *Shanghai Agricultural Science and Technology*, no. 2 (2017): 32–33.

Xu, Junwei, Meng Tianyao, and Jing Peipei, et al. "Effects of Machine-Planting Density on the Resistance and Yield of Different Types of Rice." *Acta Agronomica Sinica* 41, no. 11 (2015): 1767–1776.

Xu, Ying, Zhou Mingyao, and Xue Yafeng. "Study on Spatial Variability and Relationship between Rice Leaf Area Index and Yield." *Journal of Agricultural Engineering* 22, no. 5 (2006): 10–14.

Xuan, Haiyan, Li Suoping, and Liu Shuqun et al. "Analysis of the Relationship between Regional Precipitation, Latitude and Longitude, and Elevation." *Gansu Science Journal*, no. 04 (2006): 26–28.

Yang, Xiaoguang, Liu Zhijuan, and Chen Fu. "The Possible Effects of Global Warming on Cropping Systems in China I. The Possible Effects of Climate Warming on Northern Limits of Cropping Systems and Crop Yields in China." *Scientia Agricultura Sinica* 43, no. 2 (2010): 329–336.

Yang, Zhiying. "Performance of Jingliangyou 534 in Machine-Transplanting Cultivation Techniques in Pucheng County." *Fujian Rice and Wheat Technology* 36, no. 2 (2018): 44–46.

Yin, Mingxuan, Tao Shishun, and Zhang Rongping, et al. "Effects of Seedling Density on Effective Number of Spikes and Adult Spike Structure in Live Hybrid Rice." *Hybrid Rice* 34, no. 1 (2019): 40–43.

Yu, Liuqing, Lu Yongliang, and Xuan Songnan, et al. *Technical Regulations for Weed Control in Paddy Field.* Beijing: China Agriculture Press, 2010.

Yuan, Longping, and Ma Guohui. *Key Technologies of Super Hybrid Rice Yield of 800 kg/mu*. Beijing: China Three Gorges Publishing House, 2006.

Yuan, Longping, and Ma Guohui. *Theory and Practice of the Modified System of Rice Intensification for Super Hybrid Rice*. Changsha: Hunan Science and Technology Press, 2005.

Yuan, Longping, Wu Xiaojin, and Liao Fuming et al. *Hybrid Rice Technology*. Beijing: China Agriculture Press, 2003.

Yuan, Longping. "Proposals for Implementing the Super Hybrid Rice 'Plant Three Produce Four' High Yield Project." *Hybrid Rice* 22, no. 4 (2007): 1–2.

Yuan, Longping. *Super Hybrid Rice: Breeding and Cultivation*. Changsha: Hunan Science and Technology Press, 2020.

Zeng, Yawen, Li Zichao, and Yang Zhongyi, et al. "Geographical Distribution and Cline Classification of Indica/Japonica Subspecies of Yunnan Local Rice Resources." *Acta Agronomica Sinica* 27, no. 01 (2001): 15–20.

Zhang, Fengzhuan, Jin Zhengxun, and Ma Guohui, et al. "Dynamic Changes of Lodging Resistance and Chemical Component Contents in Culm and Sheaths of Japonica Rice During Grain Filling." *Chinese Journal of Rice Science*, no. 03 (2010): 264–270.

Zhang, Hongcheng, and Wang Fuyu. "Advances in Rice Population Research in China." *Chinese Journal of Rice Science* 15, no. 1 (2001): 51–56.

Zhang, Hongcheng, Wu Guicheng, and Wu Wenge, et al. "The SOI Model of Quantitative Cultivation of Super-High- Yielding Rice." *Scientia Agricultural Sinica* 43, no. 13 (2010): 2645–2660.

Zhao, Jin, Yang Xiaoguang, and Liu Zhijuan, et al. "The Possible Effects of Global Warming on Cropping Systems Boundary in China II. The Characteristics of Climatic Variables and the Possible Effect on Northern Limits of Cropping Systems in South China." *Scientia Agricultura Sinica* 42, no. 9 (2010): 1860–1867.

Zhou, Bin, Liu Yang, and Huang Sidi, et al. "Occurrence and Control of Rice Stemborer in Southern Hunan." *Hunan Agricultural Sciences*, no. 12 (2017): 81–84.

Zhou, Jing. "Study on Climate Ecological Adaptability and Yield Differences in Super Hybrid Indica Rice." PhD diss., Changsha: Hunan Agricultural University, 2018.

Zhu, Defeng, and Chen Huizhe. "Development of Rice Machine-Transplanting and Food Security." *China Rice* 15, no. 6 (2009): 4–7.

Zou, Yingbin. "Development of Cultivation Techniques for Double-Cropping Rice in the Yangtze River Basin." *Chinese Agricultural Sciences* 44, no. 2 (2011): 254–262.

Index

W

— ABOUT THE EDITORS —

Yuan Longping was an academician of the Chinese Academy of Engineering who worked at the Hunan Hybrid Rice Research Center. As a world-famous scientist engaged in the research of hybrid rice breeding and cultivation, he invented the three-line system for indica hybrid rice and successfully developed the two-line system. He established the technical system for super hybrid rice and proposed and implemented a high-yielding project—Planting Three to Produce Four Yield-Increasing Program of Super Hybrid Rice, with the great technical achievements he made in super hybrid rice.

Ma Guohui is a senior Hunan Hybrid Rice Research Center research fellow. As a famous agricultural expert engaged in researching high-yielding and high-efficiency cultivation of hybrid rice, he has presided over and participated in more than 20 major national scientific research projects. He has created technical achievements and systems for the high-yielding cultivation of hybrid rice and double-cropping rice cultivation under a three-cropping system.